The
Robotics
Primer

The Robotics Primer

Maja J Matarić

Illustrations by Nathan Koenig

The MIT Press
Cambridge, Massachusetts
London, England

Brief Contents

Contents

Preface

Upon arriving at USC in 1997 as an assistant professor in Computer Science, I designed the *Introduction to Robotics* course (http://www-scf.usc.edu/~csci445). The highlights of the course were the intensive hands-on LEGO robotics lab, team projects, and the end-of-semester contest held at the nearby California Science Center. I posted my lecture notes on the web, for students' convenience, and found that, over the years, a growing number of faculty and K-12 teachers around the world contacted me about using those in their courses and getting additional materials from me. In 2001, shortly after receiving tenure, being promoted to rank of associate professor, and expecting our second child, perhaps influence of the euphoria of it all, I had a deceptively simple thought: why not turn all those course notes into a book?

As I started converting my sometimes-cryptic lecture notes into book chapters, I realized the true magnitude of the challenge I had set up for myself, that of writing a textbook for an audience that intended to span K-12 and university educators and students, as well as robotics enthusiasts who wanted to go beyond the popular press and get under the surface of the topic. The readers of the book would (hopefully) span the ages of early adolescence to retirement, and education levels of elementary school to PhD. This was not going to be easy.

My motivation came from the fact that robotics is a wonderful field of study. I am not a typical engineer; instead of making robots out of old watches and radios in my parent's basement as a kid, I was interested in art and fashion design. I discovered the field late in college, through extra-curricular reading while majoring in computer science; back then there were no robotics courses at the University of Kansas. Upon entering graduate school at MIT, I serendipitously chose robotics as my direction of study, based mostly on the charisma of my PhD advisor, Rodney Brooks, who was the first to make a compelling case for it. The goal of this book is to make a

1 *What Is a Robot? Defining Robotics*

Welcome to *The Robotics Primer*! Congratulations, you have chosen a very cool way to learn about a very cool topic: the art, science, and engineering of robotics. You are about to embark on a (hopefully fun) ride that will end with your being able to tell what is real and what is not real in movies and articles, impress friends with cool facts about robots and animals, and much more, most important of which is being able to better build and/or program your own robot(s). Let's get started!

What is a robot?

This is a good question to ask because, as with any interesting area of science and technology, there is a lot of misunderstanding about what robots are and are not, what they were and were not, and what they may or may not become in the future. The definition of what a robot is, has been evolving over time, as research has made strides and technology has advanced. In this chapter, we will learn what a modern robot is.

The word "robot" was popularized by the Czech playwright Karel Capek (pronounced Kha-rel Cha-pek) in his 1921 play *Rossum's Universal Robots (R.U.R.)*. Most dictionaries list Capek as the inventor of the word "robot," but more informal sources (such as the Web) say that it was actually his brother, Josef, who made up the term. In either case, the word "robot" resulted from combining the Czech words *rabota*, meaning "obligatory work" and *robotnik*, meaning "serf." Most robots today are indeed performing such obligatory work, in the form of repetitive and fixed tasks, such as automobile assembly and DNA sequencing. However, robotics is about much more than obligatory labor, as you will see.

The idea of a robot, or some type of machine that can help people, is much older than the Capek brothers. It is not possible to pin point where it orig-

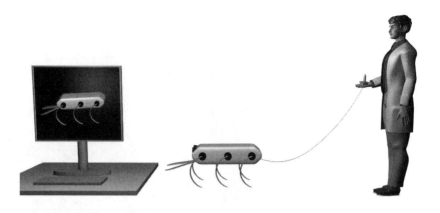

Figure 1.2 Examples of non-robots. On the left is a system that does not exist in the physical world; on the right, a system that is not autonomous. These robot wannabes are not the real thing.

SENSORS   Sensing the environment means the robot has *sensors*, some means of perceiving (e.g., hearing, touching, seeing, smelling, etc.) in order to get information from the world. A simulated robot, in contrast, can just be given the information or knowledge, as if by magic. A true robot can sense its world only through its sensors, just as people and other animals do. Thus, if a system does not sense but is magically given information, we do not consider it a true robot. Furthermore, if a system does not sense or get information, then it is not a robot, because it cannot respond to what goes on around it.

> A robot is an autonomous system which exists in the physical world, can sense its environment, and can ACT ON IT

Taking actions to respond to sensory inputs and to achieve what is desired is a necessary part of being a robot. A machine that does not act (i.e., does not move, does not affect the world by doing/changing something) is not a robot. As we will see, action in the world comes in very different forms, and that is one reason why the field of robotics is so broad.

> A *robot* is an autonomous system which exists in the physical world, can sense its environment, and can act on it to ACHIEVE SOME GOALS.

Now we finally come to the intelligence, or at least the usefulness, of a robot. A system or machine that exists in the physical world and senses it, but acts randomly or uselessly, is not much of a robot, because it does not use the sensed information and its ability to act in order to do something useful for itself and/or others. Thus, we expect a true robot to have one or more goals and to act so as to achieve those goals. Goals can be very simple, such as "Don't get stuck" or quite complex, such as "Do whatever it takes to keep your owner safe."

Having defined what a robot is, we can now define what robotics is.

ROBOTICS

Robotics is the study of robots, which means it is the study of their autonomous and purposeful sensing and acting in the physical world.

It is said that Isaac Asimov, the amazingly productive science fiction writer, was the first to use the term *robotics*, based on Capek's term *robot*. If so, he is the one who officially gave the name to the large and rapidly growing area of science and technology that this book aims to introduce.

To Summarize

- Robotics is a fast-growing field whose definition has been evolving over time, along with the field itself.

- Robotics involves autonomy, sensing, action, and achieving goals, all in the physical world.

Food for Thought

- What else can you do from afar, by means of teleoperation? You can talk, write, and see, as in telephone, telegraph, and television. There is more. Can you think of it?

- Is a thermostat a robot?

- Is a toaster a robot?

- Some intelligent programs, also called software agents, such as Web crawlers, are called "softbots." Are they robots?

- Is HAL, from the movie *2001, the Space Odyssey*, a robot?

Looking for More?

- This book comes with a free robot programming workbook you can download from the World Wide Web at:
  http://roboticsprimer.sourceforge.net/workbook/
  The workbook follows the structure of this textbook with exercises and solutions you can use to learn much more about robotics by trying it hands-on.

- Here is a fun, short little introductory book about robotics: *How to Build a Robot* by Clive Gifford. It's very easy, has great cartoons, and touches on a lot of concepts that you can then learn about in detail in this book.

- *Robotics, The Marriage of Computers and Machines* by Ellen Thro is another short introductory book you might enjoy.

- For a truly comprehensive review of modern robotics, complete with a great many robot photos, see *Autonomous Robots* by George Bekey.

- To learn more about various topics in this book and in general, look up things on Wikipedia, an ever-growing free encyclopedia found on the Web at http://en.wikipedia.org/wiki/.

2 *Where Do Robots Come From? A Brief but Gripping History of Robotics*

Have you wondered what the first robot ever built was like, who built it, when it was built, and what it did? It turns out that it is not easy to answer these questions precisely. Many machines have been built that could be called robots, depending on how one defines a robot. Fortunately, we have a modern definition of a robot from Chapter 1 and, using that definition, the first robot is typically considered to be W. Grey Walter's *Tortoise*. In this chapter, we will learn about the Tortoise, its creator, the related fields of control theory, cybernetics, and artificial intelligence (AI), and how together they played key parts in the history of robotics, from its humble origins to its current state of the art.

2.1 Control Theory

CONTROL THEORY *Control theory* is the mathematical study of the properties of automated control systems ranging from steam engines to airplanes, and much in between.

Control theory is one of the foundations of engineering; it studies a vast variety of mechanical systems that are a part of our everyday lives. Its formalisms help us understand the fundamental concepts which govern all mechanical systems. Since both the art of building automated systems and the science of studying how they work date back to ancient times, control theory can be said to have originated with the ancient Greeks, if not before. It was largely studied as part of mechanical engineering, the branch of engineering that focuses on designing and building machines, and on their physical properties, and was used to study and further develop the control of water systems (for irrigation and house-hold use), then temperature systems (for metalsmithing, animal care, etc.), then windmills, and eventually steam engines (the machines that were part of the industrial revolution). By the start of the twentieth century, classical mathematics, such as differential equations, was applied to formalizing the understanding of such systems, giving birth to what we think of as control theory today. The mathematics got more complicated as mechanisms began to incorporate electrical and electronic components.

In Chapter 10, we will discuss *feedback control*, a concept from control theory that plays an important role in robotics.

2.2   Cybernetics

While control theory was growing and maturing, another field relevant to robotics emerged. It was pioneered in the 1940s, the years around the second World War, by Norbert Wiener. Wiener originally studied control theory, and then got interested in using its principles to better understand not just artificial but also biological systems.

CYBERNETICS   The field of study was named *cybernetics*, and its proponents studied biological systems, from the level of neurons (nerve cells) to the level of behavior, and tried to implement similar principles in simple robots, using methods from control theory. Thus, cybernetics was a study and comparison of communication and control processes in biological and artificial systems.

Cybernetics combined theories and principles from neuroscience and biology with those from engineering, with the goal of finding common properties and principles in animals and machines. As we will see, W. Grey Walter's Tortoise is an excellent example of this approach.

The term "cybernetics" comes from the Greek word "kybernetes" which means governor or steersman, the name for a central component of James Watt's steam engine, which was designed based on the ideas of windmill control. In cybernetics, the idea was that machines would use a similar steersman to produce sophisticated behavior similar to that found in nature.

A key concept of cybernetics focuses on the coupling, combining, and interaction between the mechanism or organism and its environment. This interaction is necessarily complex, as we will see, and is difficult to describe formally. Yet this was the goal of cybernetics, and is still an important component of robotics. It also led to the development of the Tortoise, which can be considered to be the first robot, based on our definition in Chapter 1.

Figure 2.1   One of the tortoise robots built by W. Grey Walter. (Photo courtesy of Dr. Owen Holland)

2.2.1   Grey Walter's Tortoise

William Grey Walter (1910-1977) was an innovative neurophysiologist interested in the way the brain works. He made several discoveries, including the theta and delta brain waves, the activity your brain produces during different stages of sleep. Besides pursuing neuroscience research, he was interested in studying brain function by building and analyzing machines with animal-like behavior.

BIOMIMETIC

Today, we call machines with properties similar to those of biological systems *biomimetic*, meaning that they imitate biological systems in some way. During his research, around the 1940s and thereafter, Grey Walter built a variety of cleverly designed machines, which he called "turtles" or "tortoises," after the turtle character in Lewis Carroll's *Alice in Wonderland*, and because of their biomimetic behavior. The best known ones were named Elmer and Elsie, loosely based on the acronyms for ELectro MEchanical Robots, and Light Sensitive. The turtles were simple robots built with three wheels in a tricycle-like design, using the front wheel for steering and the two back wheels for driving. They were covered with a clear plastic shell, giving them a life-like appearance, at least to a friendly and open-minded observer. Figure 2.1 shows what one of the turtles looked like.

W. Grey Walter (the first name, William, is rarely spelled out) gave his turtles Latin names to describe their behaviors, such as Machina Speculatrix and Machina Docilis. Machina Speculatrix means "machine that thinks" or "speculates," and Machina Docilis means "machine that can be tamed/taught," by which he meant a machine that can learn, because it could be trained with whistles.

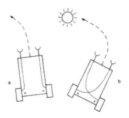

Figure 2.2 An example of Braitenberg's vehicles, in simulation. (Photo courtesy of Valention Braitenberg, "Vehicles: Experiments in Synthetic Psychology", published by The MIT Press)

BRAITENBERG'S VEHICLES

Braitenberg's vehicles started out with a single motor and a single light sensor, and gradually progressed to more motors and more sensors, and more interesting connections between them, all using analog electronic circuits. The sensors were directly connected to the motors, so that the sensory input could drive the motor output. For example, a light sensor could be connected directly to the wheels, so the stronger the light, the faster the robot moved, as if it were attracted to the light; that's called *photophilic* in Latin, literally meaning "loving light." Alternatively, in some vehicles the connection was inverted, so that the stronger the light, the slower the robot would move, as if it were repulsed by the light or afraid of it; that's called *photophobic* in Latin, literally meaning "afraid of light."

PHOTOPHILIC

PHOTOPHOBIC

EXCITATORY CONNECTION

INHIBITORY CONNECTION

A connection between the sensors and motors in which the stronger the sensory input, the stronger the motor output, is called an *excitatory connection*, because the input excites the output. Conversely, a connection in which the stronger the sensory input, the weaker the motor output, is called an *inhibitory connection*, because the input inhibits the output. The inspiration for these connections comes from biology, since they are similar to (but crude when compared with) neurons, which can have excitatory and inhibitory connections with each other. By varying the connections and their strengths (as with neurons in brains), numerous behaviors result, ranging from seeking and avoiding light, much like Grey Walter's turtles, to what looks like social behavior, and even aggression and love.

Braitenberg's book describes how such simple mechanisms can be employed to store information, build a memory, and even achieve robot learning. Some of his simpler designs have been successfully built by hobbyists and fledgling roboticists (like yourself), and have also been a source of inspiration for more advanced robotics research. Like Grey Walter's turtles, Braitenberg's vehicles were also reactive robots. We will learn more about them, too, in Chapter 14.

While the field of cybernetics concerned itself with the behavior of robots and their interaction with the environment, the new emerging field of artificial intelligence focused on, well, intelligence, naturally (no, artificially!).

2.3 Artificial Intelligence

Doesn't AI sound cool? Or, alternatively, it does it seem scary as it is usually portrayed in movies? In any case, chances are that it seems much more powerful and complex than it really is. So let's dispel the myths and see what AI has to do with robotics.

ARTIFICIAL INTELLIGENCE

The field of *artificial intelligence (AI)* was officially "born" in 1956 at a conference held at Dartmouth University, in Hanover, New Hampshire. This meeting brought together the most prominent researchers of the day, including Marvin Minsky, John McCarthy, Allan Newell, and Herbert Simon, names you can't miss if you read about the origins of AI, who are now considered founders of that field. The goal of the meeting was to discuss creating intelligence in machines. The conclusions of the meeting can be summarized as follows: in order for machines to be intelligent, they will have to do some heavy-duty thinking; and to do that, they will need to use the following:

- Internal models of the world

- Search through possible solutions

- Planning and reasoning to solve problems

- Symbolic representation of information

- Hierarchical system organization

- Sequential program execution.

Don't worry if much of the above list does not make sense right now. We will learn more about these concepts in Chapters 12 and 13, and will see that some of them may not play a fundamental role in aspects of robotics. The important outcome of the Dartmouth conference was that as the field of AI was being funded and shaped, it had a strong influence on robotics, or at least on the branch we can call *AI-inspired robotics*.

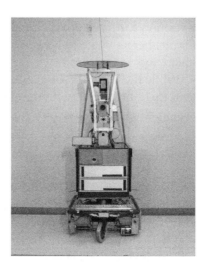

Figure 2.3 Shakey, in its specially designed world. (Photo courtesy of SRI International, Menlo Park)

SHAKEY   So what is AI-inspired robotics? *Shakey*, shown in figure 2.3, is a good example of an early AI-inspired robot. Built at the Stanford Research Institute in Palo Alto, California, in the late 1960s, it used contact sensors and a camera. As we will see in Chapter 9, cameras, which are vision sensors, provide a great deal of information to a robot and are very complicated to deal with. Because the new field of AI focused on reasoning, and vision required complex processing, those early roboticists were able to apply early AI techniques to their robots. Shakey "lived" in a very special indoor world consisting of a flat white floor and some large black objects, such as balls and pyramids. It carefully (and slowly, given the state of computer technology at that time) created plans for which way to move in that special world. People stayed out of its way or did not move around it. Given the state of robotics technology at that time, Shakey shook a bit when it tried to execute its plans: hence its name.

Shakey is a popular and well-known (now retired) early robot. Its successor was called Flakey, and you can probably guess why. Let us consider some other examples of early AI-inspired robots:

- HILARE: developed at the LAAS Lab (which stands for Laboratoire d'Analyse et D'Architecture des Systemes, loosely the "analysis and architec-

Figure 2.4 HILARE, another early robot. (Copyright LAAS-CNRS)

tures of systems") in Toulouse, France in the late 1970s, used a video camera, ultrasound sensors, and a laser rangefinder (we will learn about those in Chapters 9). Unlike most other research robots, HILARE was used by many generations of researchers and is one of the longest-lived robots to date.

- CART: developed at Stanford University in Palo Alto, California, in 1977 by Hans Moravec, as part of his doctoral thesis (just a part of it; it takes a lot to get a PhD in robotics, but it's worth it). It was literally a cart on bicycle wheels. Moravec is now considered one of the founders of modern robotics, and his recent research has been on ultrasound sensing. But back in the days of his doctoral thesis, he focused on vision sensors (so to speak), and in particular on how to use vision to move the robot VISION-BASED around, which we call *vision-based robot navigation*. (We'll talk about nav-NAVIGATION igation and its challenges in Chapter 19.) CART used vision to move very slowly, not because it was slow-moving by design, but because it was slow-thinking. The thinking took a long time, as with other early AI-inspired robots (Shakey, HILARE), because of the difficulty of processing data from vision cameras and the slowness of computer processors in those days. It is said that one can see the shadows move as time passes while CART is thinking about and planning what to do next.

- Rover: developed at Carnegie Mellon University (known as CMU), in Pittsburgh, Pennsylvania, in 1983 also by Hans Moravec. This was the robot he built after he got is doctoral degree and became a professor at CMU. Rover used a camera and ultrasound sensing for navigating. (You may be starting to notice that these old robots did nothing else but navigate, and that was hard enough. Chapter 19 will explain why that is still a hard, but much better-handled problem.) Rover was more advanced in form than CART; it looked more like modern mobile robots do these days, not like the CART did, but its thinking and acting were still very slow.

Since they thought too hard and too slowly, and failed to move very effectively and responsively, these early robots of the 1970s and 1980s provided important lessons for the fledgling field of robotics. The lessons, not surprisingly, mostly had to do with the need to move faster and more robustly, and to think in a way that would allow such action. In response, in the 1980s robotics entered a phase of very rapid development. In that decade, several new robotics directions emerged, shook things up, and eventually got orga-

TYPES OF ROBOT CONTROL

nized into the *types of robot control* that we use today: *reactive control, hybrid control, and behavior-based control*. We will learn about them in Chapters 14, 15 and 16. These are all effective replacements for what early AI-inspired robotics used, and what we now call *purely deliberative control*. We will learn about it, and why it is not in use in today's robotics, in Chapter 13.

If you are really paying attention, you have probably noticed that these early AI-inspired robots were quite different in every way from Grey Walter's tortoises and from Breitenberg's vehicles. AI was (and is) very different from cybernetics in its goals and approaches. But robotics needs to combine both, in order to create robust, fast, and intelligent machines. What it takes to do that, and how it is done, is what the rest of this book is about.

Recapping a brief but gripping history of robotics, it can be said that modern robotics emerged from the historic developments and interactions among a handful of key research areas: control theory, cybernetics, and artificial intelligence. These fields have had important and permanent impact on robotics today, and at least two of them still continue as major areas of research in their own right. Control theory thrives as a field of study applied to various machines (though typically not intelligent ones); control-theoretic ideas are at the core of low-level robot control, used for getting around. Cybernetics no longer exists under that name, but research into biologically inspired biomimetic methods for robot control is very much alive and thriving, and is being applied to mobile robots as well as humanoids. Finally, artificial intelligence is now a very large and diverse field of research that focuses largely, but not exclusively, on non physical, unembodied cognition (this basically means "disembodied thinking"; we will talk about how the body impacts thinking in Chapter 16). Robotics is growing by leaps and bounds. See

Chapter 22 to find out where all the robots are today, what they are up to, and where the field is going next.

To Summarize

- Robotics grew out of the fields of control theory, cybernetics, and AI.

- The first modern robot, W. Grey Walter's tortoise (one of them; there were several), was constructed and controlled using the principles of cybernetics.

- Braitenberg's vehicles provided more examples of cybernetics and biomimetic principles.

- Early AI influenced early AI-inspired robotics which produced Shakey, the Stanford CART, HILARE, and the CMU Rover, among others.

- AI-inspired early robots were very different from the tortoise and from Braitenberg's vehicles, and in general from biomimetic cybernetic machines.

- The current approaches to robot control, including reactive, hybrid, and behavior-based control, emerged from the influences of and lessons learned from control theory (the mathematics of controlling machines), cybernetics (the integration of sensing, action, and the environment), and AI (the mechanisms for planning and reasoning).

Food for Thought

How important is it for robots to be inspired by biological systems? Some people argue that biology is our best and only model for robotics. Others say that biology teaches us valuable lessons, but that engineering can come up with its own, different solutions. The airplanes is a popular example of non-biomimetic engineering; people's first attempts to build flying machines were inspired by birds, but today's planes and helicopters share little in common with birds. Does it matter what kind of robot you are building (biomimetic or not)? Does it matter if it is going to interact with people? For more on this last topic, wait for Chapter 22.

Looking for More?

- The two original sources on cybernetics, written by the field's founders, are: *Cybernetics or Control and Communication in the Animal and the Machine* by Norbert Wiener (1948) and *Introduction to Cybernetics* by W. R. Ashby (1956).

- W. Grey Walter wrote a book and several articles about the brain. He also published two *Scientific American* articles about his robotics ideas "An imitation of life", in 1950 (182(5): 42-45), and "A machine that learns", in 1951 (185(2): 60-63).

- Hans Moravec wrote a great many articles about robotics. He also wrote two books: *Mind Children* and *Robot: Mere Machine to Transcendent Mind*. Both books make predictions about the future of robotics and artificial life; they are so forward-looking that they sound quite like science fiction.

3 *What's in a Robot? Robot Components*

Have you wondered what makes up a robot, what makes it tick? (Actually, it better not tick; things don't tick these days, unless something is wrong!) In this chapter, we will look inside a robot to see what it is made of, in terms of both hardware and software. Peering inside will cause us to introduce and briefly review many of the concepts which will be talked about in more detail in the chapters to come. Here we will provide an overview of how these many ideas and concepts come together to constitute a robot.

Let's go back to our trusty definition of a robot:

> A robot is an autonomous system which exists in the physical world, can sense its environment, and can act on it to achieve some goals.

The definition already gives us some hints as to what a robot consists of. Specifically, it tells us that a robot's main components are:

- A physical body, so it can exist and do work in the physical world

- Sensors, so it can sense/perceive its environment

- Effectors and actuators, so it can take actions

- A controller, so it can be autonomous.

Figure 3.1 shows these required components of a real robot, and the way in which they interact with each other and with the robot's environment. Let's talk a bit about each of these components: embodiment, sensing, action, and autonomy.

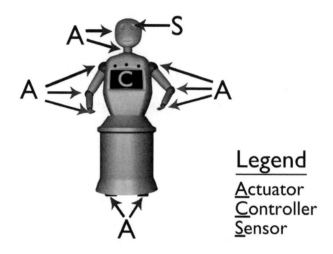

Legend

<u>A</u>ctuator
<u>C</u>ontroller
<u>S</u>ensor

Figure 3.1   The components of a robot.

3.1   Embodiment

Having a body is the first requirement for being a robot; it allows the robot to do things: move and shake, go places, meet people, and get its work done. A computer agent, no matter how realistically animated, is not a robot because it does not share the physical universe with the rest of us physical creatures; EMBODIMENT   it is not *physically embodied* in the real world. *Embodiment* refers to having a physical body. It is necessary for a robot, but it has a price:

- *We are all in the same boat:* An embodied robot must obey the same physical laws that all physical creatures obey. It can't be in more than one place at a time; it can't change shape and size arbitrarily (although some robots can do a bit of shape-shifting, as you'll see in Chapter 22); it must use the effectors on its body to actively move itself around (more on that in Chapter 4), it needs some source of energy to sense, think, and move, and when moving, it takes some time to speed up and slow down; its movement affects its environment in various ways; it cannot be invisible; and so on.

- *You are in my space:* Having a body means having to be aware of other bodies and objects around. All physical robots have to worry about not running into or hitting other things in their environment and, in some cases, about not hitting themselves (as you will see in Chapter 6). This seems easy, but it's not, and so it's typically the first thing you'll program a robot to do: avoid collisions.

- *I can only do so much:* Every body has its limitations. The shape of the robot's body, for example, has a lot to do with how it can move, what it can sense (because sensors are attached to the body, in some way or another), what work it can do, and how it can interact with other robots, cars, lawnmowers, cats, stuff, and people in its environment.

- *All in good time:* Besides obviously influencing things that have to do with space and movement, the body also influences things that have to do with time. The body determines how fast the robot can move (on its own power, at least) and react to its environment. It has often been said that faster robots look smarter; that's another example of how the robot's embodiment influences its "image."

3.2 Sensing

SENSORS   *Sensors* are physical devices that enable a robot to perceive its physical environment in order to get information about itself and its surroundings. The
SENSING   terms *sensing* and *perception* are treated as synonyms in robotics; both refer to
PERCEPTION   the process of receiving information about the world through sensors.

What does a robot need to sense? That depends on what the robot needs to do, on what its task is. A good robot designer and programmer puts the right kinds of sensors on the robot that allow it to perceive the information it needs to do its job, to achieve its goals. Similarly, animals have evolved sensors that
NICHE   are well suited for their *niche*, namely their environment and position in their ecosystem. Robots have a niche as well, consisting of their environment and their task, and just like animals, the better they fit that niche, the longer they will survive (in the lab, as a commercial product, etc.). We will talk more about this in Chapter 7.

STATE   Sensing allows the robot to know its state. *State* is a general notion from physics that is borrowed by robotics (and by computer science and AI, among other fields) and refers to the description of a system. A robot's state is a description of itself at any point in time. The robot (system) is said to be "in a state" (kind of like being "in a state of disrepair"). The more detailed the

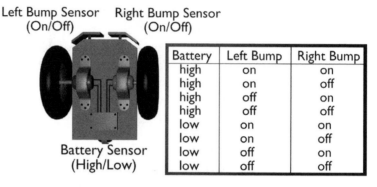

Left Bump Sensor (On/Off) Right Bump Sensor (On/Off)

Battery	Left Bump	Right Bump
high	on	on
high	on	off
high	off	on
high	off	off
low	on	on
low	on	off
low	off	on
low	off	off

Battery Sensor (High/Low)

State = (Internal State) + (External State)
     = (Battery Sensor) +
       (Left Bump + Right Bump)

Figure 3.2   A robot's sensors and its state space.

description is, the larger the state is, because it takes more bits or symbols to write it down.

To a robot (actually, to any perceiving creature), state may be visible (formally called *observable*), partially hidden (*partially observable*), or *hidden* (unobservable). This means a robot may know very little or a great deal about itself and its world. For example, if the robot cannot "feel" or "see" one of its arms, the state of that arm is hidden from it. This can be a problem if the robot needs to use the arm to reach for something, such as the hospital patient it is trying to help. At the same time, the robot may have detailed information about the hospital patient, the patient's state, through a machine that senses and monitors the patient's vital signs. We will talk more about state and what to do with it in Chapter 12.

OBSERVABLE, PARTIALLY OBSERVABLE, AND HIDDEN STATE

State may be *discrete* (up, down, blue, red) or *continuous* (3.785 miles per hour). This refers to the type and amount of information used to describe the system.

DISCRETE AND CONTINUOUS STATE

State space consists of all possible states a system can be in. For example, if a light switch can be only on or off, its state space consists of two discrete

STATE SPACE

states, on and off, and so is of size 2. If, on the other hand, the light switch has a dimmer on it, and can be turned to a variety of light levels, then it has many more states (possibly infinitely many continuous states). The term SPACE *space* in this context refers to all possible values or variations of something. Figure 3.2 shows an example of a robot that has two on/off bumpers and a high/low battery, and the state space that results from those.

It is often useful for a robot to distinguish two types of state relative to EXTERNAL AND INTERNAL STATE itself: external and internal. *External state* refers to the state of the world as the robot can perceive it, while *internal state* refers to the state of the robot as the robot can perceive it. For example, a robot may sense that it is dark and bumpy around it (external state) and also that its battery level is low (internal state).

Internal state can be used to remember information about the world (a REPRESENTATION path through a maze, a map, etc.). This is called a *representation* or an *internal model*. Representations and models have a lot to do with how complex a robot's brain is; we will talk about this at length in Chapter 12.

In general, how intelligent the robot appears to be strongly depends on how much and how quickly it can sense its environment and itself: its external and internal state.

What can a robot sense? That depends on what sensors the robot has. All of the robot's sensors, put together, create the space of all possible sensory SENSOR SPACE readings, which is called the robot's *sensor space* (also called *perceptual space*). PERCEPTUAL SPACE

The sensor or perceptual space for a robot that has just a single on-off contact sensor has two possible values: on and off. Now suppose the robot has two such sensors; in that case its sensor space consists of four values: on+on, on+off, off+on, off+off. That's the basic idea, and you can see that a robot's sensor space grows pretty quickly as you add more sensors or more complex sensors. That's one of the reasons that robots need brains.

Sensors are the machine versions of eyes, ears, noses, tongues, hairs, and various other sense organs in animals. But, as you will learn in Chapters 8 and 9, robot sensors, which include cameras, sonars, lasers, and switches, among others, are quite different from biological sensors. A robot designer and programmer needs to put her/his mind into the robot's sensor space in order to imagine how the robot perceives the world, and then how the robot should react to it. This is not easy to do, precisely because while robots share our physical world, they perceive it very differently than we do.

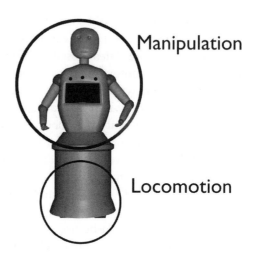

Manipulation

Locomotion

Figure 3.3 A robot that combines mobility and manipulation.

3.3 Action

EFFECTORS

Effectors enable a robot to take action, to do physical things. They are the next best thing to legs, flippers, wings, and various other body parts that allow animals to move. Effectors use underlying mechanisms, such as muscles and motors, which are called *actuators* and which do the actual work for the robot. As with sensors, robotic effectors and actuators are very different from biological ones. They are used for two main activities:

ACTUATORS

1. *Locomotion*: moving around, going places

2. *Manipulation*: handling objects.

These correspond to the two major subfields of robotics:

1. Mobile robotics; which is concerned with robots that move around, mostly on the ground, but also in the air and under water.

2. Manipulator robotics; which is concerned mostly with robotic arms of various types.

Mobile robots use locomotion mechanisms such as wheels, tracks, or legs, and usually move on the ground. Swimming and flying robots are also mobile robots, but they usually move in higher dimensions (not just on the ground), and are therefore even harder to control. Manipulators refer to various robot arms and grippers; they can move in one or more dimensions. The dimensions in which a manipulator can move are called its *degrees of freedom (DOF)*. We will learn more about that in Chapter 6.

DEGREES OF FREEDOM

The separation between mobile and manipulator robotics is slowly disappearing as more complex robots, such as humanoids which have to both move around and manipulate objects, are being developed. Figure 3.3 shows a robot that combines mobility and manipulation.

3.4 Brains and Brawn

Robots are different from animals in that biological brains take a great deal of energy (power) relative to the rest of the body, especially in humans. In robots, it's the other way around, with the actuators taking more power than the processor running the controller, the brain.

When it comes to computation, on the other hand, animals and robots are similar; both need a brain in order to function properly. For example, cats can walk without using their brain, purely using the signals from their spinal cord, and people can survive for years in a coma. Similarly, well-designed robots may move about randomly when their processor is reset. In all cases, this is not a good thing, so make sure you properly protect your brain and your robot's brain, too.

Although we won't talk much in this book about power issues, power is a major problem in practical robotics. Power issues include:

- Providing enough power for a robot without loading it down with heavy batteries

- Effectively keeping the electronics isolated from the sensors and effectors

- Preventing loss of performance as the power levels drop because batteries wear down or there is a sudden high power demand (due to running into the wall, as discussed in Chapter 4, or to a high-current sensor, as discussed in Chapter 9)

- Replenishing power in an autonomous fashion, by the robot itself instead of by people

- And many others.

As you dive into robotics, keep in mind that the brains and the brawn are directly connected and related, and should be designed and treated together.

3.5   Autonomy

CONTROLLERS *Controllers* provide the hardware and/or software that makes the robot autonomous by using the sensor inputs and any other information (for instance, whatever may be in the memory) to decide what to do (what actions to take), and then to control the effectors to execute that action. Controllers play the role of the brain and the nervous system. We say "controllers" and not "a controller" on purpose; typically there is more than one, so that various parts of the robot can be processed at the same time. For example, various sensors may have separate controllers, as may various effectors. Do all things come together in a central place? This turns out to be a major issue in robotics: they may or may not, as they may or may not in biology. We will talk more about that when we get to robot control in Chapters 13, 14, 15, and 16. Whatever kind they may be, controllers enable robots to be autonomous.

PSEUDOCODE You will see examples of controllers throughout this book. They will be written in *pseudocode* (from the Greek *pseudo* meaning "false" and *code* meaning "program"), an intuitive way of describing the controller, not any particular robot programming language.

AUTONOMY *Autonomy* is the ability to make one's own decisions and act on them. For robots, autonomy means that decisions are made and executed by the robot itself, not by human operators. Autonomy can be complete or partial. Completely autonomous robots, like animals, make their own decisions and act on them. In contrast, teleoperated robots are partially or completely controlled by a human and are more like complex puppets than like animals. According to our definition, puppetlike robots that are not autonomous at all are not really robots.

To Summarize

- The key components of a robot are sensors, effectors, and controllers.

- Sensors provide information about the world and the robot itself. They define the sensory or perceptual space of the robot and allow it to know its state, which may be discrete, continuous, observable, partially observable, or hidden.

- Effectors and actuators provide the ability to take actions. They may provide locomotion or manipulation, depending on the type of robot: mobile or manipulator.

- Controllers provide autonomy, which may be partial or complete.

Food for Thought

- What do you think is more difficult, manipulation or mobility? Think about how those abilities develop in babies, children, and then adults.

- How large do you think your sensor space is?

- Can you think of things or information in your life that is observable, partially observable, or hidden?

Looking for More?

The Robotics Primer Workbook exercises for this chapter are found here: http://roboticsprimer.sourceforge.net/workbook/Robot_Components

4 *Arms, Legs, Wheels, Tracks, and What Really Drives Them*
Effectors and Actuators

It should come as no surprise that while robotic bodies are inspired by biological ones, they are quite different from their biological counterparts in the way they are constructed and the way they work. In this chapter, we will learn about the components of a robot that enable it to take actions in order to achieve its goals.

EFFECTOR

An *effector* is any device on a robot that has an effect (impact or influence) on the environment.

Effectors range from legs and wheels to arms and fingers. The robot's controller sends commands to the robot's effectors to produce the desired effect on the environment, based on the robot's task. You can think of effectors as equivalent to biological legs, arms, fingers, and even tongues, body parts that can "do physical work" of some kind, from walking to talking. Just as sensors must be well matched to the robot's task, so must the effectors.

ACTUATOR

All effectors have some mechanism that allows them to do their work. An *actuator* is the mechanism that enables the effector to execute an action or movement.

In animals, muscles and tendons are the actuators that make the arms and legs and the backs do their jobs. In robots, actuators include electric motors, hydraulic or pneumatic cylinders, temperature-sensitive or chemically-sensitive materials, and various other technologies. These mechanisms actuate the wheels, tracks, arms, grippers, and all other effectors on robots.

Figure 4.1 A passive walker: a robot that uses gravity and clever mechanics to balance and walk without any motors. (Photo by Jan van Frankenhuyzen, courtesy of Delft University of Technology)

4.1 Active vs. Passive Actuation

PASSIVE ACTUATION

In all cases, the action of actuators and effectors requires some form of energy to provide power. Some clever designs use *passive actuation*, utilizing potential energy in the mechanics of the effector and its interaction with the environment instead of active power consumption.

Consider, for example, flying squirrels. They don't actually fly at all, but *glide* through the air very effectively, using the skin flaps between their front leggs and torso. Glider planes and hang gliders use the same principle. Birds use their wings for gliding as well, but since wings can be (and usually are) used for flying, which involves active flapping, we don't think of bird wings as a very good example of a purely passive effector. Some plants have small wings on their seeds which enable them to glide far away from the parent plant in order to spread. Nature is full of examples of such effectors that use clever designs to conserve energy and serve multiple uses, including the inspiring of roboticists.

An example of an inspired robotics design for a passive effector, originally developed by Tad McGeer, is shown in figure 4.1. McGeer developed a walking machine, called the Passive Walker, which had two legs with knees that looked basically like a pair of human legs, but had no motors or muscles or any active actuators at all. When put at the top of a downward-sloping ramp, the Passive Walker would walk in a manner that looked very humanlike.

What do you think made it walk? It was actuated by gravity, and would eventually come to rest at the bottom of the ramp. Of course, if put on a flat surface, or worse yet at the bottom of an incline, it would not walk at all, but simply fell over. It was not perfect, but it was a clever design that inspired many people. You'll notice that simple clever solutions always inspire people, not just in robotics but in any field of pursuit.

4.2 Types of Actuators

So, we have learned that there are passive and active actuators, and that there are various clever designs for both. Now, let's look into some options for active actuators. As you might imagine, there are many different ways of actuating a robotic effector. They include:

- Electric motors: These are the most common robotics actuators, the most affordable and the simplest to use, based on electric current. We'll learn much more about them in this chapter.

- Hydraulics: These actuators are based on fluid pressure; as the pressure changes, the actuator moves. They are quite powerful and precise, but are also large, potentially dangerous, must be well-packaged, and of course must be kept from leaking!

- Pneumatics: These actuators are based on air pressure; as the pressure changes, the actuator moves. Much like hydraulic actuators, they are typically large, very powerful, potentially dangerous, and must also be kept from leaking.

- Photo-reactive materials: Some materials perform physical work in response to the amount of light around them; they are called photo-reactive. Usually the amount of generated work (and thus movement) is quite small, and therefore this type of actuator is currently used only for very small, micro-scale robots. (Such photo-reactive materials are also used in tinted glasses that automatically darken as needed. However, in those cases they do not produce *movement*, and are therefore not actuators.)

- Chemically reactive materials: As the name implies, these materials react to certain chemicals. A good example is a particular type of fiber which contracts (shrinks) when put in an acidic solution and elongates (gets longer) when put in a basic (alkaline) solution. Such materials can be used as *linear actuators*; they provide linear movement (getting longer or shorter), which is very different from the rotary movement provided by motors.

LINEAR ACTUATOR

- Thermally reactive materials: These materials react to changes in temperature.

- Piezoelectric materials: These materials, usually types of crystals, create electric charges when pushed/pressed.

This is not a complete list; engineers are developing new types of actuators all the time. But let's get back to the basics and look at the simplest and most popular robotic actuators, the ones you are most likely to use: motors.

4.3   Motors

Motors are the most common actuators in robotics. They are well suited for actuating wheels, since they provide rotational movement, thereby enabling wheels to turn, and wheels, of course, are very popular effectors (in robotics and on the ground in general). Motors are also very useful for driving other types of effectors, besides wheels, as you will see next.

4.3.1   Direct-Current (DC) Motors

DC MOTOR

Compared with all other types of actuators, *direct current (DC) motors* are simple, inexpensive, easy to use, and easy to find. They can be purchased in a great variety of sizes and packages, to accommodate different robots and tasks. This is important, since you remember from earlier in this book that a good robot designer matches all parts of the robot, including the actuators, to the task.

You may know from physics that DC motors convert electrical energy into mechanical energy. They use magnets, loops of wire, and current to generate magnetic fields whose interaction turns the motor shaft. In this way, electromagnetic energy becomes kinetic energy producing movement. Figure 4.2 shows a typical DC motor; the components are hidden in the packaging, but you can see the shaft on the left and the power wires on the right.

Figure 4.2 A standard direct-current (DC) motor, the most common actuator used in robotics.

To make a motor run, you need to provide it with electrical power in the right voltage range. If the voltage is low, but not too low, the motor will still run, but will have less power. On the other hand, if the voltage is too high, the power of the motor is increased, but the wear and tear will make it break down sooner. It is much like revving a car engine; the more you do it, the sooner the engine will die.

When a DC motor is provided with nice constant voltage in the right range, it draws current in the amount proportional to the work it is doing. Work, as defined in physics, is the product of force and distance, so when a robot is pushing against a wall, the motors driving the wheels are drawing more current, and draining more of its batteries, than when the robot is moving freely, without obstacles in its way. The reason for the higher current is the physical resistance of the stubborn wall, which keeps the distance traveled small or zero, resulting in high force for the same amount of work being done. If the resistance is very high (the wall just won't move, no matter how much the robot pushes against it), the motor draws the maximum amount of power it can, and then, having run out of options, stalls. A car also stalls under similar conditions, but it's best not to try this at home.

The more current the motor uses, the more torque (rotational force) is produced at the motor shaft. This is important, since the amount of power a motor can generate is proportional to its torque. The amount of power is also proportional to the rotational velocity of the shaft. To be precise, the amount of power is proportional to the product of those two values, the torque and the rotational velocity.

When the motor is spinning freely, with nothing attached to its shaft, then its rotational velocity is the highest but the torque is zero. The output power, then, is also 0. In contrast, when the motor is stalled, its torque is the maximum it can produce but the rotational velocity is zero, so the output power is again zero. Between these extremes of free spinning and stalling, the motor actually does useful work and drives things efficiently.

How efficiently? you might ask. That depends on the quality of the motor. Some are very efficient, while others waste up to half of their energy. It gets worse with other types of motors, such as electrostatic micro-motors, which are used for miniature robots.

How fast do motors turn?

Most DC motors have unloaded (free-spinning) speeds in the range of 3000 to 9000 revolutions per minute (rmp), which is the same as 50 to 150 revolutions per second (rps). That means they produce high speed but low torque, and are therefore well suited for driving light things that rotate very fast, such as fan blades. But how often does a robot need to drive something like that? Unfortunately, not often. Robots need to do work: pull the loads of their bodies, turn their wheels, and lift their manipulators, all of which have significant mass. This requires more torque and less speed than off-the-shelf DC motors provide. So what can we do to make standard DC motors useful to robots?

4.3.2   Gearing

GEARS   We can use gears! Combining different *gears* is used to change the force and torque output of motors.

The force generated at the edge of a gear is the ratio of the torque to the radius of the gear. By combining gears with different radii, we can manipulate the amount of force and torque that gets generated.

Suppose that we have two gears, one of which is attached to the shaft of the motor and is called the input gear, the other being the output gear. The

Figure 4.3 An example of a 3-to-1 (3:1) gear reduction.

torque generated at the output gear is proportional to the torque on the input gear and the ratio of the radii of the two gears. Here is the general rule:

> *If the output gear is larger than the input gear, the torque increases. If the output gear is smaller than the input gear, the torque decreases.*

The torque is not the only thing that changes when gears are combined; there is also a corresponding change in speed. If the circumference of the input gear is twice that of the output gear then the output gear must turn twice for each rotation of the input gear in order to keep up, since the two are physically connected by means of their teeth. Here is the general rule:

> *If the output gear is larger than the input gear, the speed decreases. If the output gear is smaller than the input gear, the speed increases.*

Another way to think about it is:

> *When a small gear drives a large one, torque is increased and speed is decreased. Analogously, when a large gear drives a small one, torque is decreased and speed is increased.*

This is how gears are used to trade off the extra speed of DC motors. This is a useful trade-off because robots rarely need the high speed but could use the additional torque.

Figure 4.4 An example of ganged gears; two 3:1 gears in series produce 9:1 gear reduction.

BACKLASH

Gears are combined by "the skin of" their teeth. Gear teeth require special design so that they mesh properly. Any looseness between meshing gears, called *backlash*, makes the gear mechanism move sloppily back and forth between the teeth, without turning the whole gear. Backlash adds error to the positioning of the gear mechanism, which is bad for the robot because it won't know exactly where it is positioned. Reducing backlash requires tight meshing between the gear teeth, but that, in turn, increases friction between the gears, which wastes energy and decreases the efficiency of the mechanism. So, as you can imagine, proper gear design and manufacturing are complicated, and small, high-precision, carefully machined gearboxes are expensive.

Let's get back to looking gears in the mouth and counting the teeth. To achieve a particular gear reduction, we combine gears of different sizes (different numbers of teeth). For example, to achieve three-to-one or (3:1) gear reduction, we mesh a small gear (say one with eight teeth) with a large one (say one with $3 \times 8 = 24$ teeth). Figure 4.3 illustrates this; the large gear is the output gear and the small gear is the input gear. As a result, we have slowed down (divided) the large gear by a factor of three and at the same time have tripled (multiplied) its torque, thereby slowing down and increasing the torque of the motor. To achieve the opposite effect, we switch the input and output gear.

GEARS IN SERIES, GANGED GEARS

Gears can be organized in a *series* or *"ganged,"* in order to multiply their effect. For example, two 3:1 gears in series result in a 9:1 reduction. This requires a clever arrangement of gears. Three 3:1 gears in series can produce a 27:1 reduction. Figure 4.4 shows an example of ganged gears with the 9:1 resulting ratio. This method of multiplying reduction is the underlying mechanism that makes DC motors useful and ubiquitous (commonplace), because they can be designed for various speeds and torques, depending on the robot's tasks.

4.3.3 Servo Motors

DC motors are great at rotating continuously in one direction. But it is often useful for a robot to move a motor, and some effector attached to it, such as an arm, to a particular position.

SERVO MOTORS

Motors that can turn their shaft to a specific position are called *servo motors* or *servos* for short, because they can servo to a particular position. Servos are used in toys a great deal, such as for adjusting steering in remote-controlled (RC) cars and wing position in RC airplanes.

Would you guess that DC and servo motors are very different? Actually, they are not. In fact, servo motors are made from DC motors by adding the following components:

- Some gear reduction, for the reasons you saw above

- A position sensor for the motor shaft, to track how much the motor is turning, and in what direction

- An electronic circuit that controls the motor, in order to tell it how much to turn and in what direction.

Servo motor operation is all about getting the shaft of the motor in the desired position. This position is somewhere along 180 degrees in either direction from the reference point. Therefore, instead of turning the possible 360 degrees, the shaft of a servo motor is usually limited to only half of that, 180 degrees.

The amount of the turn within the zero to 180 degree range is specified by an electronic signal. The signal is produced as a series of pulses; when the pulse arrives, the shaft of the motor turns, and when no pulse arrives, it

WAVEFORM

PULSE-WIDTH
MODULATION

POSITION CONTROL

TORQUE CONTROL

stops. The up-and-down pattern of the pulses produces a wave like pattern, called the *waveform*.

The amount the motor shaft turns when the pulse arrives is determined by the duration of the pulse; the longer the pulse, the larger the turn angle. This is called *pulse-width modulation* because the width (duration) of the pulse modulates the signal. The exact width of the pulse is quite important, and must not be sloppy. There are no milliseconds or even microseconds to be wasted here, or the motor will behave badly, jitter, or try to turn beyond its mechanical limit (those 180-degree stops). In contrast, the duration between the pulses is not as important, since that is when the shaft is stopped.

Notice that both kinds of motors we have talked about so far, continuous rotation and servo, control the position of the shaft. Most robot actuators use *position control*, in which the motor is driven so as to track the desired position at all times. This makes motor-driven actuators very accurate, but also very *stiff*. In practice this means the actuator locks onto the desired position and, if disturbed, produces a large force in order to keep from losing that position. This can be a bad thing for a robot. One alternative is to use *torque control* instead of position control: the motor is driven so as to track the desired torque at all times, regardless of its specific shaft position. The result is an actuator that is much less stiff.

Considering the work they must do, motors, both DC and servo, require more power (more current) to run than electronics do. For comparison, consider that the 68HC11 processor, an 8-bit microprocessor used to control simple robots, draws 5 milliamps of current, while a small DC motor used on the same type of robots draws from 100 milliamps to 1 amp of current. Typically, specialized circuitry is required for control of motors. Since this is not an electronics textbook, we won't go into more detail on this topic.

4.4 Degrees of Freedom

Now that we have spent some time talking about motors as the most popular actuators, let's get back to effectors and how they move. The wheel is a pretty simple effector, whose job is to turn. A wheel can (typically) move in only one way, by turning. (As we will see later, there are some cleverly embellished wheel designs that provide sideways motion as well, but let's stick with simple turning for now.) Wheels can turn and motors make them turn.

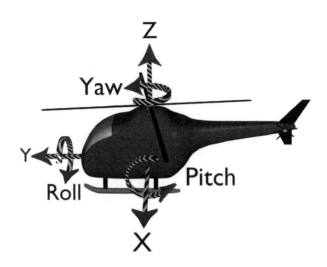

Figure 4.5 The six degrees of freedom (DOF) of a freely moving object (in this case a helicopter) in space are x, y, z, roll, pitch, and yaw.

DEGREE OF FREEDOM

A *degree of freedom* (DOF) is any of the minimum number of coordinates required to completely specify the motion of a mechanical system. You can think of it informally as a way in which the system (robot) can move.

How many degrees of freedom a robot has, is important in determining how it can impact its world, and therefore if, and how well, it can accomplish its task.

TRANSLATIONAL DOF

In general, a free (unattached, not tied, bolted, or otherwise constrained) body (such as a robot) in 3D space (normal space around you has three dimensions) has a total of six DOF. Three of those are called *translational DOF*, as they allow the body to translate, meaning move without turning (rotation). They are usually labeled x, y and z, by convention. The other three are called *rotational DOF*, as they allow the body to rotate (turn). They are called *roll, pitch, and yaw*.

ROTATIONAL DOF

ROLL, PITCH, AND YAW

As an example, imagine a flying helicopter, such as the one shown in figure 4.5. Climbing up, diving down, and moving sideways correspond to the translational DOF. Rolling from side to side, pitching up and down, and yawing (turning) left or right correspond to the rotational DOF. The six DOF together correspond to all the possible ways the helicopter could ever move. While it can't move in more than six DOF, it can move in fewer, as it turns

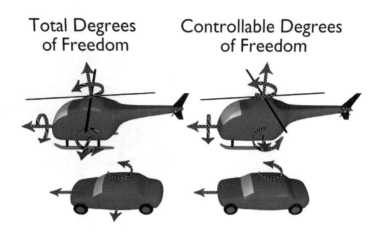

Total Degrees of Freedom   **Controllable Degrees of Freedom**

Figure 4.6   The degrees of freedom of cars and helicopters.

out. If you put the helicopter on the ground, and drive it there like a car, it has fewer DOF. How many does it have? Hang on, we'll get there.

An effector can have any number of DOF, starting with one. (Zero DOF does not make much sense, as it can't have an effect on the environment.) A wheel has one DOF, as we saw, and it usually has one motor associated with it, for actuating and controlling that DOF. But as you saw above, an effector may be passive, which means it may have DOF that are not actively actuated and controllable.

Most simple actuators, such as motors, control a single motion (up-down, left-right, in-out, etc.) of an effector. The shaft on a motor, when attached to a wheel (with or without all the gears in between), controls a single DOF, which is enough for a wheel, which has only one. But more complex effectors, such as robotic arms, have many more DOF (you'll soon see how many more), and thus require more actuators.

CONTROLLABLE DOF

UNCONTROLLABLE DOF

If a robot has an actuator for every DOF, then all of the DOF are *controllable*. This is the ideal scenario, but it is not always the case. The DOF that are not controllable are aptly called *uncontrollable DOF*. Let's work with a concrete example see what that's all about.

We agreed earlier that a helicopter has six DOF and moves in three dimensions (3D). If we were to make it move in only two dimensions (2D), on the ground, it would have fewer DOF. Consider the car, which moves on the ground in 2D only, at least when moving safely and controllably. A car has three DOF: position (x,y) and orientation (theta). This is because any body that moves on a surface (such as a car on the road, a boat on the water, or a person on the street) moves in 2D. In that flat 2D world, only three of the six DOF are possible. In the flat plane, which is 2D, a body can translate along two of the three translational DOF, but not the third, since it is the vertical dimension, out of the plane. Similarly, the body can rotate in one dimension (yaw, also called turn in 2D) but not the other two (roll and pitch), since they are outside of the plane. So by staying on the ground (on the plane), we go from 3D to 2D and from six DOF down to three.

But it gets worse. A car, like any other body in 2D, can potentially move in three ways, but it can only do so if it has effectors and actuators that can control those three DOF. And cars do not! The driver of the car can control only two things: the forward/reverse direction (through the gas pedal and the forward/reverse gear) and the rotation (through the steering wheel). So, although a car has three DOF, only two of them are controllable. Since there are more DOF than are controllable, there are motions that cannot be done with a car, such as moving sideways. That's why parallel parking is hard; the car has to be moved to the desired location between two cars and next to the curb, but the only way to get it there is through a series of incremental back-and-forth moves, instead of one sideways move that would be possible with an additional DOF. Figure 4.6 illustrates the degrees of freedom of cars and helicopters.

Before you suddenly lose faith in cars, note that the two controllable DOF of a car are quite effective: they can get the car to any position and orientation in 2D, which is about as much as you can ask for, *but* the car may have to TRAJECTORY follow a very complicated path (also called *trajectory*) to get there, and may have to stop and back up (as in parallel parking). In robotics, a formal way to say it is that a car can get to any place in the plane by following a *continuous trajectory* (meaning it does not have to fly) but with *discontinuous velocity* (meaning it has to stop and go).

In summary, uncontrollable DOF create problems for the controller in that they make movement more complicated. The ratio of the controllable DOF (let's call them CDOF) to total DOF (let's call them TDOF) on a robot tells us quite a bit about how easy it is to control its movement. There are three possibilities:

1. *CDOF = TDOF*   When the total number of controllable DOF is equal to the total number of DOF on a robot (or actuator), the ratio is 1, and the robot is said to be *holonomic*. A *holonomic* robot or actuator can control all of its DOF.

HOLONOMIC

2. *CDOF < TDOF*   When the number of controllable DOF is smaller than the total number of DOF, the ratio is less than 1, and the robot is said to be *nonholonomic*. A *nonholonomic* robot or actuator has more DOF than it can control.

NONHOLONOMIC

3. *CDOF > TDOF*   When the number of controllable DOF is larger than the total DOF, the ratio is larger than 1, and the robot is said to be *redundant*. A *redundant* robot or actuator has more ways of control than the DOF it has to control.

REDUNDANT

What does it all mean?

A good example of a holonomic mechanism is a helicopter, which has six DOF, and can have all of those six DOF controlled. The same is true for autonomous helicopters, which are robots that can fly under their own control.

A good example of a nonholonomic mechanism is of course the car, as we saw above. The ratio of controllable to total DOF in a car is 2/3, less than 1, and so we are stuck with the challenges of parallel parking. The same will be true for autonomous cars, unless they have some fancy wheels.

Can you imagine what a redundant mechanism is like?

A great example is your arm. The human arm, not including the hand, has seven DOF: three in the shoulder (up-down, side-to-side, rotation about the axis of the arm), one in the elbow (open-close), and three in the wrist (up-down, side-to-side, and again rotation). The shoulder is a little tricky, since it is based on a *ball-and-socket joint*, which is quite complicated and still something that has not been replicated in robotics. Also, the rotational DOF in the wrist really comes from the muscles and ligaments in the forearm. In spite of all that complication, in the end the human arm has seven DOF. If you have fewer, go see a doctor. If you have more, that is worthy of scientific attention. (Being double-jointed, as some people are in the elbow for example, does not increase the total number of DOF; it just increases the range of motion in the particular DOF, such as the elbow.)

BALL-AND-SOCKET
JOINT

All of the seven DOF in the human arm are controllable, as you could see while you were counting them. That brings us to the puzzle: an object in 3D

Figure 4.7 A humanlike seven DOF arm and an exoskeleton for teleoperating it. Both are made by Sarcos. (Photo courtesy of Sarcos Inc.)

space can have at most six DOF, but your arm has seven! How can your arm have more? Well, it doesn't really; the parts of the arm can still move only in 3D and within the six possible DOF, but the arm as a whole, because of the DOF in its joints, allows for more than one way to move its parts (the wrist, the fingertip, etc.) to a particular position in 3D space. This is what makes it *redundant*, meaning that there are many (infinitely many, in fact) solutions to the problem of how to move from one place to another. This is, it turns out, why the control of complicated robotic arms, such as humanoid arms, is a very hard problem that requires some rather fancy math, as discussed (but not described in detail) in Chapter 6.

Figure 4.7 shows an example of a humanlike seven DOF robot arm, and an exoskeleton that can be used to teleoperate it. The exoskeleton is called the "master arm" and the robot arm is called the "slave arm," because the former can be used to teleoperate the latter. However, the robot arm can also be programmed to be completely autonomous, and is then no slave at all, but a proper robot.

In general, effectors are used for two basic purposes:

1. Locomotion: to move the robot around

2. Manipulation: to move other objects around.

We will learn about locomotion and manipulation in the next two chapters.

To Summarize

- Effectors and actuators work together to enable the robot to do its job. Both are inspired by biological systems, but are quite different from them.

- There are several types of robot actuators, with motors being the most popular.

- Gears are used to decrease motor speed and increase power, using simple relationships of gear size, speed, and torque.

- Servo motors have particular properties that complement DC motors and are useful in both toys and robots, among other places.

- Degrees of freedom (DOF) specify how a body, and thus a robot, can move. The match between the robot's DOF and its actuators (which determine which DOF can be controlled) determines whether the robot is holonomic, nonholonomic, or redundant, and has a most important impact on what the robot can do.

Food for Thought

- How would you measure a motor's torque? What about its speed? These measurements are important for controlling robot movement. You'll see how in Chapter 7.

- How many DOF are there in the human hand?

Looking for More?

- The Robotics Primer Workbook exercises for this chapter are found here: http://roboticsprimer.sourceforge.net/workbook/Robot_Components

- *The Art of Electronics* by Paul Horowitz and Winfield Hill, is the definitive source for electronics for all, from tinkerers to researchers.

- Read Fred Martin's *Robotic Explorations: A Hands-on Introduction to Engineering* to learn more about motors, gears, and electronics, as well as ways of putting those components together with LEGOs to create interesting and fun robots.

- Sarcos creates the Disney animatronic characters, various robots you have seen in movies, the Bellagio fountains in Las Vegas, and other sophisticated robots.

- For a nice look at dimensionality, read the classic *Flatland: A Romance of Many Dimensions* by Edwin A. Abbott.

5 *Move It! Locomotion*

LOCOMOTION   *Locomotion* refers to the way a body, in our case a robot, moves from place to place. The term comes from the Latin *locus* for place and the English *motion* for movement.

You might be surprised to know that moving around presents all kinds of challenges. In fact, it is so hard that, in nature, movement requires a significant increase in "brain power." That is why creatures that move, and therefore have to contend with avoiding falling over/down/under, collisions, and being run over, and often are chasing prey and/or running away from predators, are typically smarter than those that stay put. Compare plants with even the simplest moving animals.

Most of this book will deal with robot brains, and you will see that moving the robot around will be the first challenge for those brains. But first, in this chapter we will talk about the bodies that make locomotion possible.

Many kinds of effectors and actuators can be used to move a robot around, including:

- Legs, for walking/crawling/climbing/jumping/hopping, etc.

- Wheels, for rolling

- Arms, for swinging/crawling/climbing

- Wings, for flying

- Flippers, for swimming.

Can you think of any others?

While most animals use legs to get around and go places, legged locomotion is a more difficult robotic problem compared with wheeled locomotion. The reasons for this include:

1. The comparatively larger number of degrees of freedom (DOF); as discussed in Chapter 4, the more DOF a robot has, the more complicated it is to control.

2. The challenge of stability; it is harder to stay stable on legs than it is on wheels, as we will discuss next.

5.1 Stability

STABILITY

STATIC STABILITY

Most robots need to be *stable*, meaning they do not wobble, lean, and fall over easily, in order to get their job done. But there are different ways of being stable. In particular, there are two kinds of stability: static and dynamic. A *statically stable* robot can stand still without falling over; it can be static and stable. This is a useful feature, but it requires the robot's body to have *enough legs or wheels* to provide sufficient static points of support to keep the robot stable.

Consider, for example, how you stand up. Did you know that you (along with all other humans) are not statically stable? That means you don't stay upright and balanced without expending some effort and some active control by your brain. As everyone knows, if you faint, you fall down. Although standing up may appear effortless, it takes a while to learn; babies take about a year to master the task. Standing involves using active control of your muscles and tendons in order to keep your body from falling over. This balancing act is largely unconscious, but certain injuries to the brain can make it difficult or impossible, which shows that active control is required.

If instead of having two legs we had three, standing would actually be much easier, as we could spread our legs out like a tripod. With four legs it would be even easier, and so on. In general, with more legs (or ground points), static stability becomes easier. Why do you think that is?

CENTER OF GRAVITY

Here is why: the *center of gravity (COG)* of any body needs to be above the area that is covered by the ground points (legs or wheels). When it is, the body is balanced and stays in the upright position; if the COG is not above the ground points, it drags the body down, and it starts to fall over. The area

POLYGON OF SUPPORT

covered by the ground points is called the *polygon of support*. It is called that because you can draw a projection of the body onto the surface and trace

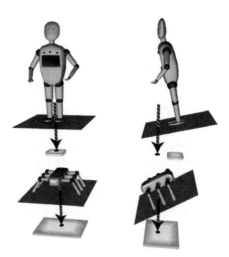

Figure 5.1 Components of balancing and stability: the COG and the polygons of support for a two-legged humanoid and a six-legged robot, shown on flat ground, and on an incline, which makes the humanoid unbalanced.

the shape or contour of the support points, which will result in a polygon. Figure 5.1 shows the COG and polygons of support for a humanoid robot and a six-legged robot, demonstrating how the humanoid is less stable.

As you can see from the figure, the polygon of support for a standing person is not very large. Our feet are relatively small compared with the rest of our bodies (if they were not, movement would be really hard, as it is when wearing huge clown shoes), and since we are relatively tall, our COG is quite high on our bodies. Keeping that COG over the relatively small polygon of support does not happen without active and trained effort. This is another reason that it takes babies a long time to learn to walk: they have very large heads compared with the rest of their bodies, and these heads bring their COG very high relative to their bodies, making balancing particularly tricky. No wonder crawling, which provides more points of support, usually comes first.

In a two-legged robot, such as a humanoid, the polygon is small, and the COG cannot be easily stably aligned to be within it and keep the robot upright. In contrast, in a three-legged robot, with its legs in a tripod organization and its body above, static stability is easy. With four legs, it's even easier, and a new feature is introduced: three legs can stay on the ground and keep the robot stable while the fourth can be lifted up, so the body can move.

> *What happens when a statically stable robot lifts a leg and tries to move?*
> *Does its COG stay within the polygon of support?*

STATICALLY STABLE
WALKING

It depends on the geometry of the robot's body and on the number of legs that stay on the ground. If the robot can walk while staying balanced at all times, we call this *statically stable walking*. A basic assumption of statically stable walking is that the weight of a leg is negligible compared with that of the body, so that the total COG of the robot is not much affected by the movement of the leg. And of course there always need to be enough legs to keep the body stable. A four-legged robot can only afford to lift only one leg at a time, since it takes at least three legs on the ground to stay statically stable. That results in a rather slow walking pace, and one that also takes a lot of energy. In fact, it is generally the case that statically stable walking, with any number of legs, while very safe, is also very slow and energy inefficient. Therefore, as a robot designer, you have to think about how important continued static stability is to your robot and if it is worth the time and effort it entails.

> *What is the alternative to static stability? Does the body have to be*
> *stable and slow, or out of control and falling over?*

DYNAMIC STABILITY

INVERSE PENDULUM

In *Dynamic stability*, the body must actively balance or move to remain stable; it is therefore called *dynamically stable*. For example, one-legged hopping robots are dynamically stable: they can hop in place or to various destinations and not fall over. These robots cannot stop and stay upright, just as you could not stop and stay standing on a pogo stick. Balancing one-legged robots (and people and other objects) is formally called the *inverse pendulum problem*, because it is the same as trying to balance a pendulum (or a stick) on one's finger, and is equally hard. Pendulums have been studied a great deal in physics and there are well-understood solutions to the problem of balancing them. Your brain solves the inverse pendulum problem whenever you stand, and so must your robot if it is dynamically stable.

Two-legged walking is only slightly easier than one-legged hopping, because it is also dynamically stable, with one leg being lifted and swung forward while the other is on the ground. Trying to get a robot to walk on two legs will really help you better understand and appreciate what your brain is doing all the time without any conscious thought.

As we have seen, people are dynamically stable walkers, while some robots are statically stable, if they have enough legs. A statically stable robot may not necessarily remain statically stable while walking. For example, no matter how many legs a robot has, if it lifts all of them, or almost all of them, it will become unstable. Therefore, a statically stable robot can use dynamically stable walking patterns, in order to be fast and efficient. In general, there is a trade-off or compromise between stability and speed of movement, as we will see next.

5.2 Moving and Gaits

GAIT A *gait* is the particular way a robot (or a legged animal) moves, including the order in which it lifts and lowers its legs and places its feet on the ground.

Desirable robot gaits have the following properties:

- Stability: the robot does not fall over

- Speed: the robot can move quickly

- Energy efficiency: the robot does not use a great deal of energy to move

- Robustness: the gait can recover from some types of failures

- Simplicity: the controller for generating the gait is not unwieldy.

Not all of the above requirements can be achieved with all robots. Sometimes safety requirements compromise energy conservation, robustness requirements compromise simplicity, and so on. As we saw in Chapter 4, it is even possible to design a robot so that it can balance and walk, in certain circumstances (usually with a push or down a hill) without any motors, as shown in figure 4.1.

Some numbers of legs and gaits are particularly popular, and can be found very commonly both in nature and in robotics. For example, most have six legs and arthropods (invertebrates with segmented bodies) have six or more (arachnids, better known as spiders, have eight), while the majority of animals have four. We humans, with our two legs, more complicated walking, and slower running, are a minority relative to most animals. We are quite slow.

Six-legged walking is highly robust and therefore common in nature, and it has served as a popular model in robotics as well. Can you guess why?

Figure 5.2   Genghis, the popular six-legged walking robot. (Photo courtesy of Dr. Rodney Brooks)

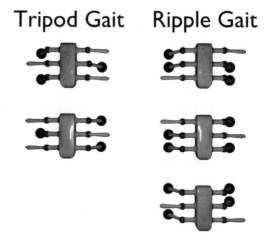

Figure 5.3   The alternating tripod gait (left) and ripple gait (right) of a six-legged walking robot. Light circles indicate feet touching the ground.

TRIPOD GAIT

ALTERNATING TRIPOD GAIT

Six legs allow for multiple walking gaits, both statically and dynamically stable. The *tripod gait* is a statically stable gait in which, as the name implies, three legs stay on the ground, forming a tripod, while the other three lift and move. If the sets of three legs are alternating, the gait is called the *alternating tripod gait* and produces quite efficient walking that can be found in various bugs (more specifically, insects and arthropods), including the common cockroach.

Numerous six-legged robots have been built. Figure 5.2 shows Genghis, a well-known simple robot that was built over only a summer by two students at the MIT Mobot Lab, part of what was then the MIT Artificial Intelligence Lab. Genghis later became a commercial product and a model for NASA rover prototypes. Its descendant Attila was a much more complicated (many many more DOF, many more sensors, a bigger brain) and less robust robot. Simplicity rules!

Almost all six-legged robots are endowed with the alternating tripod gait, because it satisfies most of the desirable gait properties listed above. In this gait, the middle leg on one side and two nonadjacent legs on the other side of the body lift and move forward at the same time, while the other three legs remain on the ground and keep the robot statically stable. Then the tripod alternates to the other side of the body, and so on.

RIPPLE GAIT   What happens if a robot has more than six legs? The alternating tripod gait can still be used, in the form of the so-called *ripple gait*, because it ripples down the length of the body. Arthropods such as centipedes and millipedes, which have many more than six legs, use the ripple gait. Figure 5.3 shows the alternating tripod and ripple gaits of a six-legged robot.

You may have noticed that we have been talking about statically stable walking even though we said earlier that such walking is not the most efficient. Indeed, when bugs run, they usually use faster, dynamically stable gaits. They may even become airborne at times, taking advantage of gliding for brief periods of time. Such extreme examples of dynamic stability demonstrate how speed and energy efficiency can be gained at the expense of stability and controller simplicity.

Controlling a robot that can balance itself on one or two legs, or can glide through the air and land safely, is a complicated control problem. Thus, not surprisingly, balance and stability are complex problems in robotics, and many robots are designed to avoid dealing with them if it is not necessary. That is why most existing mobile robots have wheels or six legs: to simplify locomotion. Robotics research, however, actively studies more complex and interesting modes of locomotion, including hopping, crawling, climbing, swimming, gliding, flying, and more.

5.3 Wheels and Steering

You might have heard the claim that wheels do not appear in nature, but that is actually not quite correct. Wheellike structures do appear in certain bacteria, but are definitely very rare in biological locomotion when compared with legs, wings, and fins. Wheels are, however, more efficient than legs, so you might wonder why animals do not have them. Evolution favors lateral symmetry, meaning it evolves bodies that have two (or more) matched sides/parts, using matched sets of genes, and so legs were easier to evolve, and they work quite well. As a result, we find them everywhere in nature; based on population size, bugs are the most abundant macro species (meaning relatively "big") and they start with six legs and go up from there. The most abundant species of all are bacteria (which are comparatively small, so micro), which dominate the living world by sheer numbers, but until micro-robotics and nano-robotics really take off (see Chapter 22), we can put those aside.

Because of their efficiency and their comparative simplicity of control, wheels are the locomotion effector of choice in robotics. Wheeled robots (as well as almost all wheeled mechanical devices, such as cars) are built to be statically stable. Although most wheels do not stray too far from the basic design, they can be constructed with as much variety and innovative flair as legs. Wheels can vary in size and shape, can use simple tires or complex tire patterns or tracks, and can even contain wheels within cylinders within other wheels spinning in different directions to provide various types of locomotion properties.

HOLONOMIC

Robots with wheels are usually designed to be statically stable, which simplifies control. However, they are not necessarily *holonomic*, meaning they cannot control all of their available degrees of freedom (DOF), as we discussed in detail in Chapter 4. In particular, most simple mobile robots have two or four wheels, and in both cases they are nonholonomic. A popular and efficient design for wheeled robots involves two wheels and a passive caster for balance, as shown in figure 5.4.

DIFFERENTIAL DRIVE
DIFFERENTIAL
STEERING

Having multiple wheels means there are multiple ways in which those wheels can be controlled. Basically, multiple wheels can move either together or independently. The ability to drive wheels separately and independently, through the use of separate motors, is called a *differential drive*. Analogously, being able to steer wheels independently is called *differential steering*.

Consider the benefits of a differential drive in the basic two-wheels-and-caster robot design. If the two wheels are driven at the same speed, the robot moves in a straight line. If one wheel (say the left) is driven faster than the other, the robot turns (in this case to the right). Finally, if the wheels are driven in the opposite directions of each other but at the same speed, the

Figure 5.4 A popular drive mechanism for simple robots, consisting of two differentially steerable driven wheels and a passive caster for balance.

robot turns in place. These abilities allow the robot to maneuver along complicated paths in order to do its job.

5.4 Staying on the Path vs. Getting There

In robot locomotion, we may be concerned with:

- Getting the robot to a particular location

- Having the robot follow a particular path (also called trajectory).

Following an arbitrary given path or trajectory is harder than having to get to a particular destination by using any path. Some paths are impossible to follow for some robots because of their holonomic constraints. For others, some paths can be followed, but only if the robot is allowed to stop, change directions (in place or otherwise), and then go again, just as we discussed with parallel parking in Chapter 4.

> *A large subarea of robotics research deals with enabling robots to follow arbitrary trajectories. Why?*

TRAJECTORY AND
MOTION PLANNING

Because there are various applications where this ability is necessary, ranging from autonomous car driving to brain surgery, and because it is not at all easy to achieve. *Trajectory planning*, also called *motion planning*, is a computationally complex process which involves searching through all possible trajectories and evaluating them, in order to find one that will satisfy the requirements. Depending on the task, it may be necessary to find the very best (shortest, safest, most efficient, etc.), so-called *optimal trajectory*. Since

OPTIMAL TRAJECTORY

robots are not just points, their geometry (shape, turning radius) and steering mechanism (holonomic properties) must be taken into account. Trajectory planning is used in mobile robots, in two dimensions, and in robot arms, in three dimensions, where the problem becomes even more complex. We talked a bit about this in Chapter 4 and will go into more detail in the next chapter.

Depending on their task, practical robots may not be so concerned with following specific trajectories as with just getting to the goal location. The ability to get to the goal is quite a different problem from planning a particular path, and is called *navigation*. We will learn more about it in Chapter 19.

To Summarize

• Moving around takes brains, or at least some nontrivial processing.

• Stability depends on the robot's geometry, and in particular the position of its COG and polygon of support. Stability can be static or dynamic.

• Static stability is safe but inefficient, dynamic stability requires computation.

• The number of legs is important. Two-legged walking is hard and slow; walking start to get easier with four legs, and much more so with six or higher.

• Alternating tripod and ripple gaits are popular when six or more legs are available.

• Wheels are not boring, and drive control is not trivial. There are many wheel designs and drive designs to choose from.

• Differential drive and steering are the preferred options in mobile robotics.

• Following a specific path or trajectory is hard, as is computing a specific path/trajectory that has particular properties (shortest, safest, etc.).

- Getting to a destination is not the same as following a specific path.

Food for Thought

- How does an automated Web path planner, such as Mapquest or Google Maps, which finds a route for, say, a trip from your house to the airport, find the optimal path?

Looking for More?

- The Robotics Primer Workbook exercises for this chapter are found here: http://roboticsprimer.sourceforge.net/workbook/Locomotion

- For some really exciting and inspiring work about insect and animal locomotion, including flying roaches and sticky geckos, see the research of Prof. Robert J. (Bob) Full at University of California, Berkeley, in his Polypedal Lab.

- For very interesting research on stick insects and the gaits they use normally or when their legs are out of order and/or when put on obstacle courses, see the work of Prof. Holk Cruse at the University of Bielefeld in Germany.

6 *Grasping at Straws*
Manipulation

In the last chapter we learned how robots get around and go places. In this chapter we will learn what they can do once they get there.

MANIPULATOR

A robotic *manipulator* is an effector. It can be any type of gripper, hand, arm, or body part that is used to affect and move objects in the environment.

MANIPULATION

Manipulation, then, refers to any goal-driven movement of any type of manipulator.

MANIPULATOR LINKS

Manipulators typically consist of one or more links connected by joints, and the endeffector. *Manipulator links* are the individual components of the manipulator which are controlled independently. If we take your arm as an example, your upper arm would be one link and your lower arm/forearm another. Links are connected by joints. We'll talk about joints a bit later in this chapter. First, let's start from the end, at the endeffector.

6.1 Endeffectors

ENDEFFECTOR

The *endeffector* is the part of the manipulator which affects the world. For example, on a hand it may be the finger that pokes, on a pointer it may be the tip that marks a spot, on a leg it may be a foot that kicks a ball. It is usually the extreme, final, end part of the manipulator, the part that is used to do the work (at least most of the time; the entire manipulator or large parts of it may also be used, see Food for Thought below for an example).

In contrast to locomotion, where the body of the robot moves itself to get to a particular position and orientation, a manipulator typically moves itself in order to get the endeffector to the desired three-dimensional (3D) position and orientation. Most traditional manipulation is concerned with the problem of bringing the endeffector to a specific point in 3D space. This is a surprisingly difficult problem. Why do you think that is?

Because the endeffector is attached to the arm, and the arm is attached to the body, manipulation involves not only moving the endeffector, but also JOINT LIMIT considering the whole body. The arm must move so as to avoid its own *joint limit* (the extreme of how far the joint can move), the body, and any other obstacles in the environment. As a result, the path followed by the hand, and thus the arm, and perhaps even the whole body, to achieve the desired task (getting the endeffector to the desired position) may be quite complex. More importantly, computing that path is a complex problem that involves thinking about space with the robot's body, manipulator, and task in mind. FREE SPACE Specifically, it involves computing the *free space* of the particular manipulator, body, and task (the space in which movement is possible), and then searching that space for paths that get to the goal.

Because all of the above is far from easy, autonomous manipulation is very challenging. The idea of autonomous manipulation was first used in teleoperation, where human operators moved artificial arms to handle hazardous materials.

6.2   Teleoperation

TELEOPERATION As we learned in Chapter 1, *teleoperation* means controlling a machine from a REMOTE CONTROL distance. Teleoperation is related to *remote control*, but the two terms are not used in the same way in robotics. In particular, teleoperation usually means controlling a complex manipulator or rover (as in teleoperated NASA rovers used for space exploration), while remote control is usually used for control of simple mechanisms, such as toy cars. The term remote control is not used in robotics, although its literal meaning is basically the same as teleoperation.

As was mentioned in Chapter 1, puppeteering is a form of teleoperation. It's no surprise that puppeteering is an art that requires a great deal of skill. Similarly, teleoperating complicated manipulators and robots is also a complex task that requires a great deal of skill. The challenge stems from the following:

- Complexity of the manipulator: The more degrees of freedom (DOF) there are in the manipulator being controlled, the harder it is to control.

- Constraints of the interface: Initially, joysticks were the most common interface between a teleoperator and the arm being controlled. But using a joystick to control a complicated mechanical arm, such as a seven DOF humanlike one, requires a great deal of training and concentration.

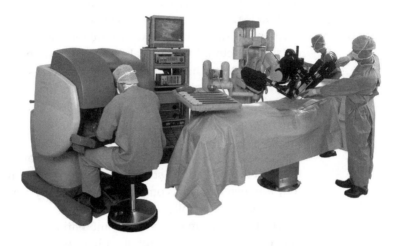

Figure 6.1 A complex teleoperated manipulator robot used in robot-assisted surgery in some hospitals. (Photo ©[2007] Intuitive Surgical, Inc.)

- Limitations of sensing: It is difficult to control an arm without directly seeing what it touches and where it moves, as well as feeling the touch and the resistance, and so on.

ROBOT-ASSISTED
SURGERY

Teleoperation has been used with great success in some very complex domains, such as *robot-assisted surgery*, where it has been applied to hip surgery, cardiac (heart) surgery, and even brain surgery. In some cases, as in cardiac surgery, the robot moves inside the body of the patient to cut and suture (crudely put, to sew up), while the surgeon controls it from outside the body with the movements of his or her fingers and hands, which are connected to wires that transmit the signal to the robot's manipulators. The major benefit of such minimally invasive surgery, made possible by teleoperation, is that instead of having to cut the patient's chest open and crack some ribs to get to the heart, only three small holes are all it takes to push in the manipulators and perform the surgery. As a result, there are fewer post-operation problems and infections, and the patient recovers much faster. Figure 6.1 shows a teleoperated robot used in robot-assisted surgery.

But even for well-designed teleoperation interfaces, it takes quite a bit of training for the human operator to learn to control the manipulator(s), for the reasons given above. One popular way of simplifying this problem (somewhat at least) is to use an exoskeleton.

EXOSKELETON   An *exoskeleton* (from the Greek *exo* meaning "outside") literally means an outside skeleton. In biological systems, it is in the form of a hard outer structure, such as the shell of an insect, that provides protection or support.

In robotics, an exoskeleton is a structure that a human wears and controls; it provides additional strength or, in teleoperation, allows for sensing the movements and the forces the human produces in order to use those to control a robot. Exoskeletons do not qualify as robots according to our definition (see Chapter 1) because they do not make their own decisions or act on them; they just provide added strength or dexterity to human operators. Therefore, they are merely shells, completely operated by a human and typically without sensors or controllers.

Teleoperation can be used in combination with autonomy, allowing a human user to influence the robot without having to control it continuously.

In summary, teleoperation is one means of simplifying manipulator control, but it does not help solve the problem of autonomous manipulator control, so let us return to that.

6.3   Why Is Manipulation Hard?

Why is manipulation so hard? As was the case in locomotion, there is typically no direct and obvious connection between what the effector needs to do in physical space and what the actuator does to move that effector in order to get the job done.

KINEMATICS   The correspondence between actuator motion and the resulting effector motion is called *kinematics*. Kinematics consist of the rules about the structure of the manipulator, describing what is attached to what.

JOINTS   The various parts (links) of a manipulator are connected by *joints*. The most common joint types are:

ROTARY JOINT   • *Rotary* (ball-and-socket), providing rotational movement around a fixed axis

PRISMATIC JOINT   • *Prismatic* (like a piston), providing linear movement.

Robot manipulators can have one or more of each type of joint. Robotic joints typically provide one controllable DOF, and in order to have maximum generality of movement, a separate actuator (typically a motor, such as the one you saw in figure 4.2 in Chapter 4) is used to move each of the DOF. This

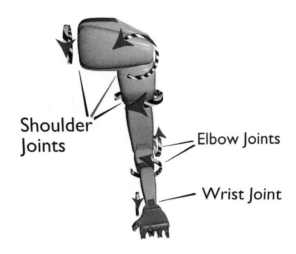

Shoulder Joints

Elbow Joints

Wrist Joint

Figure 6.2 An anthropomorphic (humanlike) robot arm.

immediately tell you that if you want to have many DOF in a manipulator, you will need a lot of physical machinery to make up the mechanism and move it in order to achieve the desired controlled movement. (And then you'll need batteries to power those motors, which are heavy, so the motors now have to be stronger to lift the manipulators that carry them, which takes power, and so on. Designing complicated robots is fun but not easy.)

Consider again the human arm (not including the hand) with its seven DOF. Note that the arm itself has only three joints: the shoulder, the elbow, and the wrist, and those three joints control the seven DOF. This means some of the joints control more than one DOF; specifically, the shoulder joint has three DOF, as does the wrist. How does that work? Looking more closely at human anatomy, we find that the shoulder joint is a ball-and-socket joint (just like the hip) and is actuated by large muscles. It turns out that ball-and-socket joints are very difficult to create in artificial systems, not only robots but also realistically modeled animated characters in computer games. The wrist, unlike the shoulder, is a smaller joint controlled by a collection of muscles and ligaments in the arm. It is also complex, but in an entirely different way from the shoulder.

Trying to get a robot to look and move like an animal (any animal, from a rat to a human) is very hard, at least in part because on robots we use motors

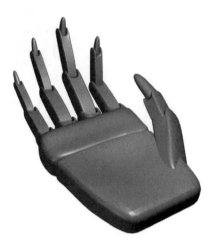

Figure 6.3   An anthropomorphic (humanlike) robot hand.

to drive mechanical links connected by rotary joints. In contrast, animals use muscles, which are linear actuators, and are comparatively much lighter, springier, more flexible, and stronger. There is ongoing research into artificial muscles, but it is in early stages and far from ready for use in robots. Most manipulator actuation is done with motors.

ANTHROPOMORPHIC   Simulating the function of an *anthropomorphic*, meaning humanlike in shape (from the Greek *anthropo* for "human," and *morph* for "shape"), manipulator in a robot is challenging. An anthropomorphic robot arm is shown in figure 6.2.

The human arm is complicated enough, but it is simple compared with the human hand, an extremely complex and versatile manipulator. The hand has a very large number of joints and, therefore, a very large number of DOF. It also has a very compact design; if you were to construct a robotic hand with the same number of DOF, you would need a lot of space to fit all the motors in. Or you could move the motors away from the hand, but they would have to go somewhere: up the arm or elsewhere on the body. An example of an anthropomorphic robot hand is shown in figure 6.3.

Humans use their *general-purpose* hands to manipulate tools, such as knives and forks and screwdrivers and hammers, for getting specific jobs done. In contrast, robot manipulators are typically *specialized*, and may already have

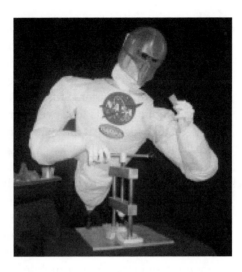

Figure 6.4 NASA's Robonaut humanoid torso robot. (Photo courtesy of NASA)

an attached tool at the endpoint, such as a welding gun, a paint sprayer, or a screwdriver. More general-purpose robots, on the other hand, may have multi-purpose grippers that allow them to grab and use various tools. Finally, some complex humanlike robotic hands have been developed for a large variety of possible tasks. NASA's Robonaut humanoid torso robot, shown in figure 22.6, is a good example, with humanlike arms and hands for fixing problems on the space shuttle and space station.

Regardless of the task at hand (pun intended), in order to control a robot manipulator we have to know its kinematics: what is attached to what, how many joints there are, how many DOF for each joint, and so on. All of these properties can be stated formally, allowing us to use mathematics in order to solve *manipulation problems*, problems about where the endpoint is relative to the rest of the arm, and how to generate paths for the manipulator to follow in order to get the job done.

MANIPULATION PROBLEMS

To move the endeffector of a manipulator to a desired point, we typically need to compute the angles for the joints of the whole manipulator. This conversion from a Cartesian (x,y,z) position of the endpoint (e.g., a fingertip) and the angles of the whole manipulator (e.g., an arm) is called *inverse kinematics*. The name refers to the fact that this is the opposite of the simpler process of figuring out where the endpoint of the manipulator is given the joint angles

INVERSE KINEMATICS

for all of the joints. That was *kinematics*, presented earlier in this chapter. The process of computing inverse kinematics is expensive (computationally intense). Why is that so? To answer that question we would need to get into some math, and it may be best to read about it in one of the sources listed at the end of this chapter. Suffice it to say that it is necessary, and important, since enabling robots to move their arms to the right place at the right time is necessary for a great many uses, from shaking hands to surgery (described earlier).

DYNAMICS

After kinematics, the next aspect of control in general, and manipulation in particular, for us to consider is dynamics. *Dynamics* refers to the properties of motion and energy of a moving object. Since robots move around and expend energy, they certainly have dynamics; and the faster they move, the more significant their dynamics, and therefore the more impact dynamics have on the robot's behavior. For example, the behavior of a slow-moving mouse-like maze-solving mobile robot is not very strongly impacted by its dynamics, while the behavior of a fast-moving tennis-ball-juggling torso robot certainly is. No surprise there.

Analogously to direct and inverse kinematics, manipulation also involves the computation of direct and inverse dynamics. These computations are even more complicated and computationally expensive. You can learn more about this rather mathematical topic from the sources listed below (and may be glad for not having the math included here).

GRASP POINTS

COMPLIANCE

Because the steps involved in robot manipulation are challenging, problems such as reaching and grasping constitute entire subareas of robotics research. These fields study topics that include finding *grasp points* (where to put the fingers relative to the center of gravity, friction, obstacles, etc.), the force/strength of grasps (how hard to squeeze so the robot neither drops nor crushes the object), *compliance* (yielding to the environment required for tasks with close contacts, such as sliding along a surface), and performing highly dynamic tasks (juggling, catching), to name just a few. Manipulation is of particular interest as robots are poised to enter human environments in order to assist people in a variety of tasks and activities. For this, they must be able to interact effectively with and manipulate a variety of objects and situations, a tall order indeed for robot designers and programmers.

To Summarize

- Robot manipulators consist of one or more links, joints connecting the links, and an endeffector.

- There are two basic types of joints (rotary and prismatic) and a variety of types of endeffectors (grippers, hands, pointers, tools, etc.).

- The process of manipulation involves trajectory planning, kinematics, and dynamics, and it is quite computationally complex.

- Teleoperation can be used for manipulation, in order to avoid autonomous control.

Food for Thought

- How many DOF are there in the human hand? Can you control each of them independently?

- Which of the two joint types we discussed, rotational and prismatic, is more commonly found in biological bodies? Can you think of examples of specific animals?

- Are astronaut suits exoskeletons? What about a lever-controlled backhoe?

- Imagine an exoskeleton which has its own sensors, makes its own decisions, and acts on them. This is definitely a robot, yet it is also controlled by a human and attached to the human's body. Can you imagine where this could be useful?

- Robotic dogs have been used to play soccer. Because they are not designed for that task, they move slowly and have trouble aiming and kicking the ball (not to mention finding it, but that's not part of manipulation; it belongs to sensing, discussed in the next few chapters). Some of the best robot soccer teams have come up with a clever idea: instead of having the dogs walk as they normally do, they lower them so that the two rear legs are driving and steering the robot, while the two front legs are on the ground, used to grasp the ball and kick it, which is much easier. This turns out to work very well, even if it looks rather silly. What are the endeffectors of the dogs in that case? Are the dogs now purely mobile robots?

Looking for More?

Here are some good resources for reading more about the topics we mentioned in this chapter:

- *Robot Modeling and Control* by Mark Spong

- *Modeling and Control of Robot Manipulators* by L. Sciavicco, B. Siciliano

- *Principles of Robot Motion: Theory, Algorithms, and Implementations* by H. Choset, K. M. Lynch, S. Hutchinson, G. Kantor, W. Burgard, L. E. Kavraki and S. Thrun

- *Planning Algorithms* by Steven M. LaValle

7 *What's Going On?*
Sensors

Knowing what is going on is a requirement for survival, not to mention for intelligent behavior. If a robot is to achieve anything at all, it must be able to sense the state of its own body (its internal state; see Chapter 3) and the state of its immediate environment (external state; also see Chapter 3). In fact, as we learned in Chapter 1, for a robot to be a robot, it must be able to sense. In this chapter you will find out how the robot's ability to sense directly influences its ability to react, achieve goals, and act intelligently.

A robot typically has two types of sensors based on the source of the information it is sensing:

PROPRIOCEPTION

1. *Proprioceptive sensors:* These sensors perceive elements of the robot's internal state, such as the positions of the wheels, the joint angles of the arms, and the direction the head is facing. The term comes from the Latin word *proprius* meaning "one's own" (also featured in "proprietor," "property," and "appropriate"). *Proprioception* is the process of sensing the state of one's own body. It applies to animals as much as it does to robots.

EXTEROCEPTION

2. *Exteroceptive sensors:* These sensors perceive elements of the state of the external world around the robot, such as light levels, distances to objects, and sound. The term comes from the Latin word *extra* meaning "outside" (also featured in "extrasensory," "extraterrestrial," and "extroverted"). *Exteroception* is the process of sensing the world around the robot (including sensing the robot itself).

PERCEPTUAL SYSTEM

Proprioceptive sensors and exteroceptive sensors together constitute the *perceptual system* of a robot, as shown in figure 7.1. However, one of the main challenges of robotics is that sensors themselves do not provide convenient *state* information to the robot. For example, sensors do not say: "There is

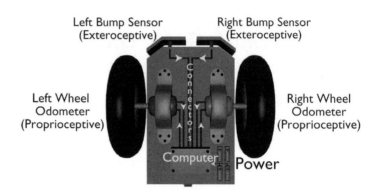

Left Bump Sensor
(Exteroceptive)

Right Bump Sensor
(Exteroceptive)

Left Wheel
Odometer
(Proprioceptive)

Right Wheel
Odometer
(Proprioceptive)

Computer   Power

Figure 7.1 Proprioceptive and exteroceptive sensors on a simple robot.

a blue chair on your left, and your grandmother Zelda is sitting in it and looking uncomfortable." Instead, sensors may tell the robot the light levels and colors in its field of view, whether it is touching something in a particular area, whether there is a sound above some threshold, or how far away the nearest object is, and so on.

Rather than being magic providers of all the information the robot could possibly need, *sensors* are physical devices that measure physical quantities. Table 7.1 considers some devices and the quantities they measure.

As table 7.1 illustrates, the same physical property may be measurable with more than one type of sensor. This is very convenient, as we will find out, since sensor information is prone to noise and errors, so acquiring information from multiple sensors can provide improved accuracy.

Sensor noise and errors, which are inherent in physical measurement and cannot be avoided, contribute to one of the major challenges of robotics: uncertainty. *Uncertainty* refers to the robot's inability to be certain, to know for sure, about the state of itself and its environment, in order to take absolutely optimal actions at all times. Uncertainty in robotics comes from a variety of sources, including:

SENSORS

UNCERTAINTY

Physical Property	$\rightarrow$	Sensing Technology
Contact	$\rightarrow$	bump, switch
Distance	$\rightarrow$	ultrasound, radar, infra red
Light level	$\rightarrow$	photocells, cameras
Sound level	$\rightarrow$	microphones
Strain	$\rightarrow$	strain gauges
Rotation	$\rightarrow$	encoders and potentiometers
Acceleration	$\rightarrow$	accelerometers and gyroscopes
Magnetism	$\rightarrow$	compasses
Smell	$\rightarrow$	chemical sensors
Temperature	$\rightarrow$	thermal, infra red
Inclination	$\rightarrow$	inclinometers, gyroscopes
Pressure	$\rightarrow$	pressure gauges
Altitude	$\rightarrow$	altimeters

Table 7.1 Some sensors and the information they measure.

- Sensor noise and errors

- Sensor limitations

- Effector and actuator noise and errors

- Hidden and partially observable state

- Lack of prior knowledge about the environment, or a dynamic and changing environment.

Fundamentally, uncertainty stems from the fact that robots are physical mechanisms that operate in the physical world, the laws of which involve unavoidable uncertainty and lack of absolute precision. Add to that imperfect sensor and effector mechanisms and the impossibility of having total and perfect knowledge, and you can see why robotics is hard: robots must survive and perform in a messy, noisy, challenging real world. Sensors are the windows into that world, and in robotics those windows are, so far, quite small and hard to see through, metaphorically speaking.

We can think of various robot sensors in terms of the amount of information they provide. For example, a basic switch is a simple sensor that provides one bit of information, on or off. *Bit*, by the way, refers to the fundamental unit of information, which has two possible values, the binary digits 0 and 1; the word comes from "b(inary) (dig)it." In contrast, a simple camera lens (a vision sensor) is stunningly rich in information. Consider a standard camera, which has a 512 X 512 pixel lens. A *pixel* is the basic element of the

BIT

PIXEL

image on the camera lens, computer, or TV screen. Each of these 262,144 pixels may, in the simplest case, be either black or white, but for most cameras it will provide much more range. In black-and-white cameras, the pixels provide a range of gray levels, and in color cameras they provide a spectrum of colors. If that sounds like a lot of information, consider the human *retina*, the part of your eye that "sees" and passes information to the brain. The retina is a light-sensitive membrane with many layers that lines the inner eyeball and is connected to the brain via the optic nerve. That amazing structure consists of more than a hundred million photosensitive elements; no wonder robots are quite a way behind biological systems.

RETINA

Although we just mentioned this, it is worth repeating:

> *Sensors do not provide state. They provide raw measured quantities, which typically need to be processed in order to be useful to a robot.*

The more information a sensor provides, the more processing is needed. Consequently, it takes no brains at all to use a switch, but it takes a great deal of brains (in fact, a large portion of the human brain) to process inputs from vision sensors.

Simple sensors that provide simple information can be employed almost directly. For example, consider a typical mobile robot that has a switch at the front of the body. When the switch is pushed, the robot stops, because it "knows" that it has hit something; when the switch is not pushed, the robot keeps moving freely. Not all sensory data can be processed and used so simply, which is why we have brains; otherwise, we would not need them.

In general, there are two ways in which sensory information can be treated:

1. We can ask the question: "Given that sensory reading, what should I do?" or

2. We can ask the question: "Given that sensory reading, what was the world like when the reading was taken?"

ACTION IN THE WORLD RECONSTRUCTION

Simple sensors can be used to answer the first question, which is about *action in the world*, but they do not provide enough information to answer the second question, which is about *reconstruction of the world*. If the switch on the robot indicates that it has hit something, that is all the robot knows; it cannot deduce anything more, such as the shape, color, size, or any other information about the object it has contacted.

At the other extreme, complex sensors such as vision provide a great deal more information (a lot of bits), but also require a great deal more processing to make that information useful. They can be used to answer both questions asked above. By providing more information, a sensor can allow us to attempt to *reconstruct* the world that produced the reading, at least with respect to the particular measured quantity. In a camera image we can look for lines, then objects, and finally try to identify a chair or even a grandmother in the image. For an idea of what it takes to do that, see Chapter 9.

SIGNAL-TO-SYMBOL

The problem of going from the output of a sensor to an intelligent response is sometimes called the *signal-to-symbol problem*. The name comes from the fact that sensors produce signals (such as voltage levels, current, resistance, etc.), while an action is usually based on a decision involving symbols. For example, the rule: "If grandmother is there and is smiling, approach her, otherwise go away" can be easily programmed or encoded with symbols for grandmother and smiling, but it is a lot more complicated to write the rule using raw sensor data. By using symbols, we can make information *abstract* and not sensor-specific. But getting from a sensory output (of any sensor) to such an abstract symbolic form (or encoding) of information in order to make intelligent decisions is a complex process. While it may seem obvious, this is one of the most fundamental and enduring challenges in robotics.

SENSOR
PREPROCESSING

Since sensors provide signals and not symbolic descriptions of the world, they must be processed in order to extract the information the robot needs. This is usually called *sensor preprocessing* because it comes before anything else can be done in terms of using the data to make decisions and/or take action. Sensor (pre)processing can be done in different ways and draws on methods from signal processing (a branch of electrical engineering) and computation.

7.1 Levels of Processing

ELECTRONICS

Suppose that your robot has a switch sensor to detect bumping into obstacles, as described above. To figure out if the switch is open or closed, you need to measure the voltage going through the circuit. That is done with *electronics*.

SIGNAL PROCESSING

Now suppose your robot has a microphone sensor for recognizing a voice. Besides the electronic processing, it will need to separate the signal from any background noise and then compare it with one or more stored voices in order to perform recognition. That is done with *signal processing*.

Next suppose your robot has a camera for finding your grandmother in the room. Besides the electronics and signal processing, it will need to find the objects in the room, then compare them against a large database in order

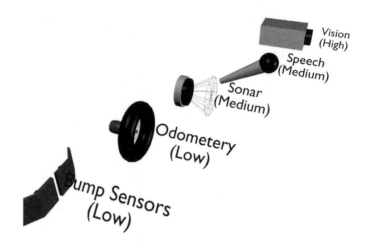

Vision
(High)

Speech
(Medium)

Sonar
(Medium)

Odometery
(Low)

Bump Sensors
(Low)

Figure 7.2   Levels of sensory processing.

COMPUTATION   to try to recognize the grandmother. That is done with *computation*.

As you can see, sensory information processing is challenging and can be computationally intensive and time consuming. Of the above processes, computation is the slowest but most general. For any specific problem, a system can be designed to solve it at a lower, and thus faster, level. For example, although computation is usually necessary for processing visual images, specialized microprocessors, so-called "vision chips" have been developed that are designed for particular vision-based tasks, such as recognizing particular faces or fruits or engine parts. They are fast but very specialized; they cannot be used to recognize anything else.

Given that a great deal of processing can be required for perception (as shown in figure 7.2), we can already see why a robot needs some type of brain. Here is what a robot needs in order to process sensory inputs:

- Analog or digital processing capabilities (i.e., a computer)

- Wires to connect everything together

- Support electronics to go with the computer

- Batteries to provide power for the whole thing.

This means that perception requires:

- Sensors (power and electronics)

- Computation (more power and electronics)

- Connectors (to connect it all).

It is generally not a good idea to separate what the robot senses, how it senses it, how it processes it, and how it uses it. If we do that, we end up with a large, bulky, and ineffective robot. Historically, perception has been studied and treated in isolation, and typically as a reconstruction problem, assuming that a robot always has to answer the second question posed above. None of these approaches has resulted in effective methods that robots can use to better sense the world and go about their business.

Instead, it is best to think about the what, why, and how of sensing as a single complete design, consisting of the following components:

- The task the robot has to perform

- The sensors best suited for the task

- The mechanical design most suited to enable the robot to get the sensory information necessary to perform the task (e.g., the body shape of the robot, the placement of the sensors, etc.).

Robotics researchers have figured out these important requirements of effective perception, and have been exploring various methods, including:

- *Action-oriented perception* (also called "active sensing"): instead of trying to reconstruct the world in order to decide what to do, the robot can use the knowledge about the task to look for particular stimuli in the environment and respond accordingly. As an example, it is very hard for robots to recognize grandmothers in general, but it is not nearly as hard to look for a particular color pattern of a grandmother's favorite dress, perhaps combined with a particular size and shape, and speed of movement.

  A clever psychologist named J. J. Gibson wrote (in 1979) about the idea that perception is naturally biased by what the animal/human needs to do (i.e., action), and is influenced by the interaction between the animal and its environment; it is not a process that retrieves absolute "truth" about the environment (i.e., reconstruction). This idea has been quite popular in action-oriented perception.

- *Expectation-based perception:* Use knowledge about the robot's environment to help guide and constrain how sensor data can be interpreted. For example, if only people can move in the given environment, we can use motion by itself to detect people. This is how simple burglar alarms work: they don't recognize burglars, or even people, just detect movement.

- *Task-driven attention:* Direct perception where more information is needed or likely to be provided. Instead of having the robot sense passively as it moves, move the robot (or at least its sensors) to sense in the direction where information is most needed or available. This seems very obvious to people, who unconsciously turn their heads to see or hear better, but most robots still use fixed cameras and microphones, missing the opportunity to perceive selectively and intelligently.

- *Perceptual classes:* Divide up (partition) the world into perceptual categories that are useful for getting the job done. That way, instead of being confronted with an unwieldy amount of information and numerous possibilities, the robot can consider manageable categories it knows how to handle. For example, instead of deciding what to do for every possible distance to an obstacle, the robot may have only three zones: too close, ok but be on the alert, and not to worry. See Chapter 14 for an example of just such a robot.

The idea that sensor function (what is being perceived) should decide sensor form (where the sensor should be, what its shape should be, if/how it should move, etc.) is employed cleverly in all biological systems. Natural evolved sensors have special geometric and mechanical properties well suited for the perceptual tasks of the creature that "wears" them. For example, flies have complex compound, faceted eyes, some birds have polarized light sensors, some bugs have horizon line sensors, humans have specially shaped ears that help us figure out where the sound is coming from, and so on. All of these, and all other biological sensors, are examples of clever mechanical designs that maximize the sensor's perceptual properties, its range and accuracy. These are very useful lessons for robot designers and programmers.

As a robot designer, you will not get the chance to make up new sensors, but you will always have the chance (and indeed the need) to design interesting ways of using the sensors you have at your disposal.

Here is an exercise: How would you detect people in an environment?

Remember the lessons from this chapter so far: Use the interaction with the world and keep in mind the task.

The obvious answer is to use a camera, but that is the least direct solution to the problem, as it involves a great deal of processing. You will learn more about this in Chapter 9. Other ways of detecting people in the environment include sensing:

- Temperature: Search for temperature ranges that correspond to human body temperature

- Movement: If everything else is static, movement means people

- Color: Look for a particular range of colors corresponding to people's skin or their clothes or uniforms

- Distance: If an otherwise open distance range becomes blocked, there is likely a moving human around.

The above are just some ways of detecting people by using sensors that are simpler than vision and require less processing. They are not perfect, but compared with vision, they are fast and easy. Often these alone, or in combination, are enough to get the task done, and even when vision sensors are available, other sensory modalities can improve their accuracy. Consider burglar alarm sensors again: they sense movement through temperature changes. While they could confuse a large dog with a human, in indoor environments canine (dog) burglars are rare, so the sensors tend to be perfectly suited to the task.

Now let's do another exercise: how would you measure distance to an object?

Here are some options:

- Ultrasound sensors provide distance measurements directly (time of flight)

- Infra red sensors can provide it through return signal intensity

- Two cameras (i.e., stereo) can be used to compute distance/depth

- A camera can compute distance/depth by using perspective (and some assumptions about the structure of the environment)

- A laser and a fixed camera can be used to triangulate distance

- A laser-based, structured light system, can overlay a grid pattern over the camera image and the processing system can use the distortions in that pattern to compute distance.

These are just some of the available means of measuring distance, and, as you saw, distance is one measure that can be used to detect other objects and people.

SENSOR FUSION   Combining multiple sensors to get better information about the world is called *sensor fusion*.

Sensor fusion is not a simple process. Consider the unavoidable fact that every sensor has some noise and inaccuracy. Combining multiple noisy and inaccurate sensors therefore results in more noise and inaccuracy and thus more uncertainty about the world. This means some clever processing has to be done to minimize the error and maximize accuracy. Furthermore, different sensors give different types of information, as you saw above in the example of detecting people. Again, clever processing is necessary to put the different types of information together in an intelligent and useful way.

As usual, nature has an excellent solution to this problem. The brain processes information from all sensory modalities - vision, touch, smell, hearing, sound - and a multitude of sensors. Eyes, ears, and nose are the obvious ones, but consider also all of your skin, the hairs on it, the strain receptors in your muscles, the stretch receptors in your stomach, and numerous others sources of body awareness (proprioception) you have at your disposal, consciously or otherwise. A great deal of our impressive brain power is involved in processing sensory information. Therefore it is not surprising that this is a challenging and important problem in robotics as well.

To Summarize

- Sensors are the robot's window into the world (through exteroception) as well as into its own body (through proprioception).

- Sensors do not provide state, but instead measure physical properties.

- Levels of sensor processing can include electronics, signal processing, and computation. Simple sensors require less processing, and thus less associated hardware and/or software.

- The what, why, and how of robot sensing, the form and the function, should be considered as a single problem.

- Nature gives us ample examples of clever sensor form and function.

- Action-oriented perception, expectation-based perception, focus of attention, perceptual classes, and sensor fusion can all be used to improve the availability and accuracy of sensor data.

- Uncertainty is a fundamental and unavoidable part of robotics.

Food for Thought

- Uncertainty is not much of a problem in computer simulations, which is why simulated robots are not very close to the real, physical ones. Can you figure out why?

- Some robotics engineers have argued that sensors are the main limiting factor in robot intelligence: if only we had more, smaller, and better sensors, we could have all kinds of amazing robots. Do you believe that is all that's missing? (Hint: If that were so, wouldn't this book be much thinner?)

- Being able to sense the self, being self-aware, is the foundation for consciousness. Scientists today still argue about what animals are conscious, and how that relates to their intelligence, because consciousness is a necessary part of higher intelligence of the kind people have. What do you think will it take to get robots to be self-aware and highly intelligent? And if some day they are both, what will their intelligence be like, similar to ours or completely different?

Looking for More?

- The Robotics Primer Workbook exercises for this chapter are found here: http://roboticsprimer.sourceforge.net/workbook/Sensors

- Jerome James Gibson (1904-1979) is considered one of the most important contributors to the field of visual perception. The ideas we discussed in the chapter come from his classic book *The Perception of the Visual World*, written in 1950, in which he put forward the idea that animals "sampled" information from their environment. In the same book he introduced the notion of "affordance," which is very important in machine vision (and also, as it happens, in ergonomic design). To learn more, look for Gibson's book or for books on machine vision. For suggestions for those, see Chapter 9.

- You can learn about sensor fusion from the work of Prof. Robyn Murphy. In fact, after reading this book, you should consider reading her *Introduction to AI Robotics*, which covers sensor fusion and many other topics at a more advanced level.

8 Switch on the Light
Simple Sensors

As we learned in Chapter 7, we can consider a sensor simple if it does not require a great deal of processing to yield useful information to the robot. In this chapter we will take a closer look at several such simple sensors, including switches, light sensors, position sensors, and potentiometers.

But first, let's consider another way in which we can categorize all sensors, both simple and complex. Sensors can be divided into two basic categories: active and passive.

8.1 Passive vs. Active Sensors

PASSIVE SENSORS
DETECTOR
ACTIVE SENSORS

Passive sensors measure a physical property from the environment. They consist of a *detector*, which perceives (detects) the property to be measured. In contrast, *active sensors* provide their own signal/stimulus (and thus typically require extra energy) and use the interaction of that signal with the environment as the property they measure. Active sensors consist of an *emitter* and

EMITTER

a *detector*. The *emitter* produces (emits) the signal, and the detector perceives (detects) it.

Passive sensors can be simple or complex. In this chapter we will learn about some simple passive sensors, including switches and resistive light sensors, and in Chapter 9 we will learn about cameras, currently the most complex passive sensors. Analogously, active sensors are not necessarily complex. In this chapter we will learn about reflectance and break beam sensors, which are simple and active, and in the next chapter we will learn about ultrasound (sonar) and laser sensors, both of which are complex and active.

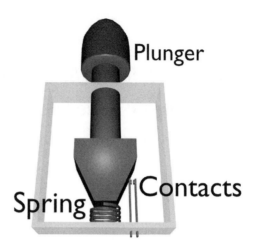

Figure 8.1   A basic switch, used for interacting with lights and robots, among other uses.

Remember, whether a sensor is complex is determined by the amount of *processing* its data require, while whether a sensors is active is determined by its *mechanism*.

Let's start by taking a look at the simplest passive sensors: switches.

8.2   Switches

Switches, such as the one shown in figure 8.1, are perhaps the simplest sensors of all. They provide useful information at the electronics (circuit) level, since they are based on the principle of an open vs. a closed circuit. If a switch is *open*, no current can flow though the circuit; if it is *closed*, current can flow through. By measuring the amount of current flowing through the circuit, we can tell if the switch is open or closed. So, switches measure the change in current resulting from a closed circuit, which in turn results from physical contact of an object with the switch.

Depending on how you wire a switch to a circuit, it can be normally open or normally closed. Either way, the measurement of current is all you need in order to use the switch as a sensor. This simple principle is applied in a wide variety of ways to create switches, and switches are, in turn, used in a variety of clever ways for sensing, such as:

CONTACT SENSORS
- *Contact sensors* detect when the sensor has contacted another object (e.g., they trigger when a robot hits a wall or grabs an object).

LIMIT SENSORS
- *Limit sensors* detect when a mechanism has moved to the end of its range (e.g., they trigger when a gripper is wide open).

SHAFT ENCODERS
- *Shaft encoder sensors* detects how many times a motor shaft turns by having a switch click (open/close) every time the shaft turns.

You use many kinds of switches in your everyday life: light switches, computer mouse buttons, keys on keyboards (computers and electronic pianos), buttons on the phone, and others.

The simplest yet extremely useful sensor for a robot is a *bump switch*, also called a *contact switch*, which tells the robot when it has bumped into something. Knowing this, the robot can back away and get around the obstacle, or stick to it, or keep bumping it, whatever is consistent with its goals. You'll discover that even for such a simple sensor as a bump switch, there are many different ways of implementation.

As we know from numerous examples in biology, building a clever body structure around a sensor can make that sensor much more sensitive and accurate. Therefore, switches can be attached to a great variety of places and components on a robot. For instance, a switch can be attached to a large, rigid (for example, plastic) surface so that when any part of the surface contacts an object, the switch closes. This is a good way to find out if any part of a robot's front, say, or side, has hit an obstacle. Another clever way to use a switch is in the form of a whisker, as found on many animals.

Can you think of how to build a (simple) whisker from the principles of a switch?

Here is one way: Attach a long wire to a switch; whenever the wire is bent enough, the switch will close and thus indicate contact. But this is not very sensitive, as the whisker has to bend quite a lot to close the switch. What can you do to fix this? Well, you can use a rigid material instead of a flexible wire for the whisker; this will make it more sensitive, and in fact more similar to the bump sensor we talked about above, for the chassis of the robot's body. But the whisker could also break if the robot does not stop. What else can you do?

A more effective way to build a whisker sensor is as follows: Use a metal (conductive) wire placed in a metal (conductive) tube. When the whisker bends, it contacts the tube, and thus closes the circuit. By adjusting the length and width of the tube, the whisker sensitivity can be tuned for the particular task and environment of the robot.

The examples above just scratch the surface of the numerous ways to cleverly design and place switches to make a robot aware of contact with objects in its environment.

Touch/contact sensors abound in biology, in much more sophisticated forms than the simple switches we have discussed so far. Whiskers and antennae are the biological inspiration for the sensors we have talked about in this section. You can consider any part of your skin to be a sensor for contact, as well as for pressure and heat, and every hair on your body as a whisker. Such abundance of sensory input is not yet available to machines; it's no wonder we are far from having truly "sensitive" robots.

8.3   Light Sensors

Besides being able to detect contact with objects, it is often useful for a robot to be able to detect areas of darkness and light in its environment. Why? Because a light can then be used to mark a special area, such as the battery recharging station or the end of a maze. With a light sensor, a robot can also find dark places and hide there.

What other uses for light sensors can you think of?

PHOTOCELL   Light sensors measure the amount of light impacting a photocell. *Photocells*, as the name indicates (*photo* means "light" in Greek), are sensitive to light, and this sensitivity is reflected in the resistance in the circuit they are attached to. The resistance of a photocell is low when it is illuminated, sensing a bright light; it is high when it is dark. In that sense, a light sensor is really a "dark" sensor. This can be made simpler and more intuitive; by simply inverting the output in the circuit, you can make it so that low means dark and high means light.

Figure 8.2 shows a photocell; the squiggly line is the photoresitive part that senses/responds to the light in the environment. Basic household night lights use the same photocells as some robots, in order to detect and respond to particular light levels. In night lights, low light causes the bulb to light up, while in the robot it may result in changing speed or direction, or some other

Figure 8.2 An example of a photocell.

appropriate action. Remember Braitenberg's vehicles in Chapter 2? They used this simple principle to produce all sorts of interesting robot behaviors, such as following, chasing, avoiding, and oscillating, which resulted in complex interpretations by observers, that included repulsion, aggression, fear, and even love.

Light sensors are simple, but they can detect a wide range of wavelenghts, much broader than the human eye can see. For example, they can be used to detect ultraviolet and infra red light, and can be tuned to be sensitive to a particular wavelength. This is very useful for designing specialized sensors; you will see how later in this chapter.

Just as we saw with switches, light sensors can be cleverly positioned, oriented, and shielded in order to improve their accuracy and range properties. They can be used as passive or active sensors in a variety of ways, and they can measure the following properties:

- Light intensity: how light/dark it is

- Differential intensity: difference between photocells

- Break in continuity: "break beam," change/drop in intensity.

We will see examples of all of those uses in this chapter. Another property of light that can be used for sensing is polarization.

8.3.1   Polarized Light

POLARIZING FILTER

"Normal" light emanating from a light source consists of light waves that travel in all directions relative to the horizon. But if we put a *polarizing filter* in front of the light source, only the light waves with the direction of the filter will pass through it and travel away. This direction is called the "characteristic plane" of the filter; "characteristic" because it is special for that filter, and

POLARIZED LIGHT

"plane" because it is planar (two-dimensional). *Polarized light* is light whose waves travel only in a particular direction, along a particular plane.

Why is this useful? Because just as we can use filters to polarize the light, we can use filters to detect light with a particular polarization. Therefore we can design *polarized light sensors*.

In fact, we don't have to limit ourselves to just one filter and thus one characteristic plane of light. We can combine polarizing filters. How does that work? Consider a light source (such as a lightbulb) covered with a polarizing filter. The resulting polarized light is only in the characteristic plane of the filter. What happens if we put another, identical filter in the path of the resulting polarized light? All of the light gets through. But what if we use a filter with a 90-degree angle of polarization instead? Now none of the light gets through, because none of it is in the characteristic plane of the second filter.

By playing around with photocells and filters and their arrangement, you can use polarized light to make specialized sensors that cleverly manipulate what and how much light is detected. These are active sensors, since they consist not only of a photocell (for detecting the light level) but also of one (or more) light source(s) (for emitting the light) and one (or more) filter(s) for polarizing the light. The general idea is that the filtering happens between the emitter and the receiver; exactly where – whether closer to the emitter, to the receiver, or both – depends on the robot, its environment, and its task.

Polarized light sensing exists in nature as well; many insects and birds use polarized light to navigate more effectively.

8.3.2   Reflective Optosensors

Reflective optosensors, as you can guess by the name, operate on the principle of reflected light. These are active sensors; they consist of an emitter and

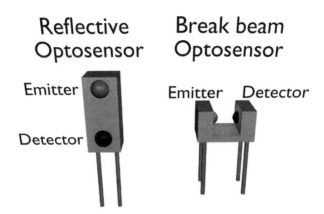

Figure 8.3 The two types of optosensor configuration: reflectance (on the left) and break beam (on the right).

LIGHT-EMITTING DIODE (LED)

a detector. The emitter is usually made with a *light-emitting diode (LED)*, and the detector is usually a photodiode/phototransistor. You can learn about the details of what these electronic components consist of as suggested in the Looking for More section at the end of this chapter.

Reflective optosensors do not use the same technology as the resistive photocells we learned about in the last section. Resistive photocells are nice and simple, but their resistive properties make them slow, because change in resistance takes time. Photodiodes and phototransistors, on the other hand, are much faster, and are therefore preferred for use in robotics.

There are two basic ways in which reflective optosensors can be arranged, based on the relative positions of the emitter and the detector:

REFLECTANCE SENSORS

1. *Reflectance sensors*: The emitter and the detector are side by side, separated by a barrier; the presence of an object is detected when the light reflects from it and back into the detector.

BREAK BEAM SENSORS

2. *Break beam sensors*: The emitter and the detector face one another; the presence of an object is detected if the beam of light between the emitter and the detector is interrupted or broken (thus the name) .

8.3.3   Reflectance Sensors

What can you do with this simple idea of measuring light reflectance/reflectivity?

Here are just a few of the many useful things you can do:

- Detect the presence of an object: Is there a wall in front of the robot?

- Detect the distance to an object: How far is the object in front of the robot?

- Detect some surface feature: Find the line on the floor (or the wall) and follow it.

- Decode a bar code: Recognize a coded object/room/location/beacon.

- Track the rotations of a wheel: Use shaft encoding; we will learn more about this later in this chapter.

Although the general idea of using reflected light is simple, the exact properties of the process are anything but. For example, light reflectivity is affected by the color, texture (smoothness or roughness), and other properties of the surface it hits. A light-colored surface reflects light better than a dark-colored one; a matte (non-shiny) black surface reflects little if any light, and so is invisible to a reflective sensor. Therefore, it may be harder and less reliable to detect dark objects using reflectance than light ones. In the case of determining the distance to an object, the same principle makes lighter objects that are far away seem closer than dark objects that are close.

This gives you part of the idea why sensing in the physical world is challenging. No sensor is perfect, and all sensors are prone to error and noise (interference from the environment). *Therefore, even though we have useful sensors, we cannot have complete and completely accurate information.* These intrinsic, unavoidable limitations of sensors are rooted in the physical properties of the sensor mechanism, and have an impact on the resulting accuracy or the sensor. Therefore, these fundamental limitations are part of *uncertainty in robotics*.

AMBIENT LIGHT

Let's talk about sensor noise as it relates to reflective light sensors. A light sensor has to operate in the presence of the light existing in the environment, which is called *ambient light*. The reflectance sensor must ignore ambient light in order to be sensitive only to its own emitter's reflected light. This is difficult if the wavelength of the ambient light is the same as that of the emitter. The sensor's mechanism has to somehow subtract or cancel out the ambient light from the detector reading, so that it can accurately measure only the light coming from the emitter, which is what it needs.

How is that done? How does the detector know the amount of ambient light?

It has to sense it. The ambient light level is measured by taking a sensor reading with the emitter *off*. To measure only the emitter's reflected light, the detector takes two (or more, for higher accuracy) readings of the sensor level, one with the emitter *on* and one with it *off*. When one is subtracted from the other (and the signs are adjusted properly so we don't end up with negative light), the difference produces the amount of the emitter's light reflected back to the sensor. This is an example of sensor calibration.

CALIBRATION *Calibration* is the process of adjusting a mechanism so as to maximize its performance (accuracy, range, etc.). Sensors require calibration, some only initially, and others continually, in order to operate effectively. Calibration may be performed by the designer, the user, or the sensor mechanism itself.

Going back to our reflective optosensor, we've just seen how to calibrate it in order to subtract ambient noise. But if we do this calibration only once, when the sensor is first used, we may find that the sensor becomes very inaccurate over time. Why is that? Because ambient light levels change during the course of the day, lights get turned on and off, bright or dark objects may be present in the area, and so on. Therefore, if the environment can change, the sensor has to be calibrated repeatedly to stay accurate and useful. There is no time to sit back and relax when it comes to sensing.

As we mentioned above, ambient light is a problem if it is the same wavelength as the emitter light, and therefore interferes with it and is hard to cancel out. The general way to avoid interference is to encode the signal from the sensor in a way that the detector can easily separate it from ambient light. One way to do this is through the use of polarization filters, as we saw earlier. Another way is through adjusting the wavelength of the emitted light. Here is how.

8.3.4 Infra Red Light

VISIBLE LIGHT *Visible light* is light in the frequency band of the electromagnetic spectrum that human eyes can perceive.[1] Infra red (IR) light has a wavelength different

---

1. If you remember the electromagnetic spectrum from physics, that helps in this chapter, but is not necessary.

from visible light and is not in the visible spectrum. IR sensors are a type of light sensors which function in the infra red part of the frequency spectrum. They are used in the same ways that visible light sensors are: as reflectance sensors or as break beams.

Both of those uses are forms of active sensors. IRs are not usually used as passive sensors in robotics. Why is that? Because robots do not usually need to detect ambient IR. But other types of sensors, such as IR goggles, better known as "night vision" glasses, do detect ambient IR passively. They collect and enhance the light in the IR spectrum and transform it into the visible spectrum. When tuned to the frequency of human body heat, they can be (and are) used to detect human movement in the dark.

IR is very useful in robotics because it can be easily modulated, and in that way made less prone to interference. Such modulated IR can also be used for communication (i.e., for transmitting messages), which is how IR modems work. So let's learn about modulation.

8.3.5 Modulation and Demodulation of Light

MODULATED LIGHT
DEMODULATOR

Light is *modulated* by rapidly turning the emitter on and off, pulsing it. The resulting pulsed signal is then detected by a *demodulator*, a mechanism that is tuned to the particular frequency of the modulation, so it can can be decoded. The detector has to sense several "on" flashes in a row in order for the demodulator to determine its frequency and decode it.

Strobe light is a type of modulated visible light, and given how hard it is to look at a strobe light, you can see (so to speaks) why visible light is not usually used in modulated form: it's too hard on the eyes. However, modulated IR is commonly used, since IR is not in the visible spectrum. Most household remote controls are based on modulated IR, including the fought-over TV channel changer.

8.3.6 Break Beam Sensors

You probably have a good intuitive idea of how a break beam sensor works. In general, any pair of compatible emitter-detector devices can be used to produce break beam sensors, including:

- An incandescent flashlight bulb and a photocell

- Red LEDs and visible-light-sensitive phototransistors

- Infra red IR emitters and detectors.

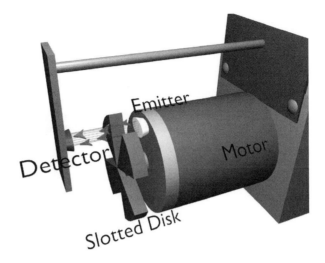

Figure 8.4 Break beam shaft encoder mechanism.

Where have you seen break beam sensors? Images from movies might come to mind, with crisscrossed laser beams and clever burglars making their way through them. More realistically, one of the most common uses of break beam sensing is not in plain sight, because it is packed inside motor mechanisms and used to keep track of shaft rotation. Here is how it works.

8.3.7 Shaft Encoders

SHAFT ENCODER *Shaft encoders* measure the angular rotation of a shaft or an axle. They provide position and/or velocity information about the shaft they are attached to. For example, the speedometer measures how fast the wheels of the car are turning, and the odometer measures the number of rotations of the wheels. Both speedometers and odometers use shaft encoding as the underlying sensing mechanism.

In order to detect a turn, or a part of a turn, we have to somehow mark the thing that is turning. This is usually done by attaching a round notched disk to the shaft. If the shaft encoder uses a switch, the switch clicks every time the shaft completes a full rotation. More commonly, a light sensor is used: a light emitter is placed on one side of the disk, and a detector on the other, in a break beam configuration. As the disk spins, the light from the emitter

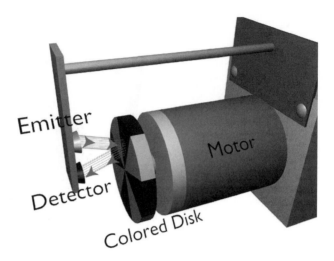

Figure 8.5   Reflectance-based shaft encoder mechanism.

reaches the detector only when the notched part of the disk passes in front of it.

If there is only one notch in the disk, then every time the notch passes between the emitter and the detector, this means the disk has completed one full rotation. This is useful, but it allows for measuring with only a low level of precision. If any noise or error is present, one or more turns might be missed, and the encoder will thus be quite inaccurate.

To make the encoder more accurate as well as more precise, many notches are cut into the disk. The break beam principle is still the same: whenever the light gets through, it is sensed by the detector and counted. Figure 8.4 shows what the mechanism looks like. You can see that it is important to have a fast sensor if the shaft turns very quickly. That is why a resistive sensor would not be appropriate; it is comparatively slow, while an optosensor works well for this propose, as we discussed earlier in this chapter. (If you forgot, just remember that optosensors use light, which travels faster than anything else.)

An alternative to cutting notches in the disk is to paint the disk with wedges of alternating, contrasting colors. The best color choices are black (absorbing, nonreflecting) and white (highly reflecting), as they provide the highest contrast and the best reflective properties. But in this case, since there are no notches in the disk, how does the break beam sensor work? It doesn't. This is not a break beam sensor, but a reflectance sensor. Instead of putting the sensor in the break beam configuration, in this case the emitter and the detector are placed on the same side of the disk, side by side, in a reflectance configuration. Figure 8.5 shows what that looks like.

Regardless of whether the shaft encoding sensor is a break beam or a reflectance sensor, the detector will output a wave function (a sequence of the signal being on and off, up and down) of the sensed light intensity of the emitter. This output is then processed, using signal processing, with hardware or a simple processor, to calculate the position and the speed by counting the peaks of the waves.

We can use encoders in at least two ways to measure the speed of a robot:

- Encode and measure the speed of a driven wheel

- Encode and measure the speed of a passive wheel (caster) that is dragged by the robot.

Why might we want to use the second of the above? One use is for robots that have legs instead of wheels. For example, the designers of Genghis, the six-legged robot mentioned in Chapter 5, had it drag a little wheel in some of the experiments, to measure the distance it had traveled, especially when it was learning how to walk (see Chapter 21 to learn about that).

We can combine the position and velocity information the encoder provides to have the robot do more sophisticated things, such as move in a straight line or turn at by exact angle. However, doing such precise movements is quite difficult, because wheels tend to slip and slide, and there is usually some backlash in the gearing mechanism (recall Chapter 4). Shaft encoders can provide feedback to correct some of the errors, but having some error remain is unavoidable. There is no perfect sensor, since uncertainty is a fact of life.

So far, we've talked about detecting position and velocity, but did not talk about direction of rotation. Suppose the robot's wheel suddenly changes the direction of rotation; it would be useful for the robot to be aware of it. If the change is intentional, the encoder can tell the robot how accurate the turn was; and if the change is unintentional, the encoder may be the first or only way the robot will know it has turned.

QUADRATURE SHAFT
ENCODING

The mechanism for detecting and measuring direction of rotation is called *quadrature shaft encoding*. One place where you might have used it is inside an old-fashioned computer mouse, the kind that has a ball inside (not the more recent, optical type). In such a mouse, the direction of rotation of the ball, and thus the direction of the movement of the mouse itself, is determined through quadrature shaft encoding.

Quadrature shaft encoding is an elaboration of the basic break beam idea: instead of using only one sensor, use two. The two encoders are aligned so that their two inputs coming from the detectors are 90 degrees (one quarter of a full circle, thus the name quadrature) out of phase. By comparing the outputs of the two encoders at each time step with the output of the previous time step, we can tell if there is a direction change. Since they are out of phase, only one of them can change its state (i.e., go from on to off or vice versa) at a time. Which one does it determines in which direction the shaft is rotating. Whenever a shaft is moving in one direction, a counter is incremented in that encoder, and when it turns in the opposite direction, the counter is decremented, thus keeping track of the overall position of the mechanism.

CARTESIAN ROBOTS

In robotics, quadrature shaft encoding is used in robot arms with complex joints, such as the ball-and-socket joints we discussed in Chapter 4. It is also used in *Cartesian robots*, which are similar in principle to Cartesian plotter printers, and are usually employed for high-precision assembly tasks. In those, an arm moves back and forth along an axis or gear.

We have seen that switches and light sensors can be used in a variety of different ways, and in some cases in the same ways (as in shaft encoding). Let's talk about one more type of simple sensor.

8.4   Resistive Position Sensors

As we just learned, photocells are resistive devices that sense resistance in response to the light. Resistance of a material, it turns out, can change in response to other physical properties besides light. One such property is tension: the resistance of some devices increases as they are bent. These passive "bend sensors" were originally developed for video game controls, but have since been adopted for other uses as well.

As you might expect, repeated bending fatigues and eventually wears out the sensor. Not surprisingly, bend sensors are much less robust than light sensors, although the two use the same underlying principle of responding to resistance.

By the way, your muscles are full of biological bend sensors. These are proprioceptive sensors that help the body be aware of its position and the work it is doing.

8.4.1 Potentiometers

Potentiometers, popularly known as "pots," are commonly used for manual tuning of analog devices: they are behind every knob or slider you use on a stereo system, volume control, or light dimmer. These days, it's getting

DIGITAL harder to find such knobs, since most devices are *digital*, meaning using discrete (from the Latin *discretus* meaning "separate") values. The word "digital" comes from the Latin *digitus* meaning finger, and most electronic devices today are tuned digitally, by pushing buttons (with fingers), and thus they use switches rather than pots. But not so long ago, tuning to radio stations and adjusting volume, among other things, were done with pots.

Potentiometers are resistive sensors; turning the knob or pushing a slider effectively alters the resistance of the sensor. The basic design of potentiometers involves a tab that slides along a slot with fixed ends. As the tab is moved, the resistance between it and each of the ends of the slot is altered, but the resistance between the two ends remains fixed.

In robotics, potentiometers are used to tune the sensitivity of sliding and rotating mechanisms, as well as to adjust the properties of other sensors. For example, a distance sensor on a robot may have a potentiometer attached to it which allows you to tune the distance and/or sensitivity of that sensor manually.

You might be thinking that these simple sensors are not so very simple after all. And you are right; sensor mechanisms can get pretty complex pretty fast, but they are not much compared with biological sensors. Still, keep in mind that while these are nontrivial physical mechanisms, the resulting sensor data are quite simple and require little processing. In the next chapter we'll move on to some complex sensors that spew out much more data and give us more processing work to do.

To Summarize

- Sensors can be classified into active and passive, simple and complex.

- Switches may be the simplest sensors, but they provide plenty of variety and have a plethora of uses, including detecting contact, limits, and turning of a shaft.

- Light sensors come in a variety of forms, frequencies, and uses, including simple photocells, reflective sensors, polarized light and infra red (IR) sensors.

- Modulation of light makes it easier to deal with ambient light and to design special-purpose sensors.

- There are various ways to set up a break beam sensor, but they are most commonly used inside motor shaft encoders.

- Resistive position sensors can detect bending and are used in a variety of analog tuning devices.

Food for Thought

- Why might you prefer a passive to an active sensor?

- Are potentiometers active or passive sensors?

- Our stomach muscles have stretch receptors, which let our brains know how stretched our stomach are, and keep us from eating endlessly. What robot sensors would you say are most similar to such stretch receptors? Are they similar in form (mechanism of how they detect) or function (what they detect)? Why might stretch receptors be useful to robots, even without stomachs and eating?

Looking for More?

- The Robotics Primer Workbook exercises for this chapter are found here: http://roboticsprimer.sourceforge.net/workbook/Sensors

- The all-around best text for learning about electronics as well as debugging hardware problems is *The Art of Electronics* by Paul Horowitz and Winfield Hill (Cambridge University Press). Every robot lab worth its beans has at least one well-worn copy.

- *Sensors for Mobile Robots: Theory and Applications* by H. R. (Bart) Everett is a comprehensive and reader-friendly textbook covering all of the sensors we have overviewed in this chapter, and quite a few more.

9 *Sonars, Lasers, and Cameras*
Complex Sensors

Congratulations, you've graduated from simple to complex sensors, and now you are in for it! The sensors we have seen so far, passive or active, do not require a great deal of processing or computation in order to provide information directly useful to a robot. However, the information they provide is itself simple and limited: light levels, presence or absence of objects, distance to objects, and so on. Not only are complex processing and computation not necessary, but they would be of little use. Not so with complex sensors. In contrast to simple ones, complex sensors provide much (much, much) more information, that can yield quite a bit more useful fodder for the robot, but also requires sophisticated processing.

In this chapter, we will learn about ultrasound, laser, and vision sensors, some of the most commonly used complex sensors in robotics. But don't assume that they are the only complex sensors available; there are others (such as radar, laser radar, GPS, etc.), and new ones are always being developed.

9.1 Ultrasonic or Sonar Sensing

ULTRASOUND *Ultrasound* literally means "beyond sound," from the Latin *ultra* for "beyond" (used here in the same way as in "ultraviolet" and "ultraconservative") and refers to a range of frequencies of sound that are beyond human
SONAR hearing. It is also called *sonar*, from so(und) na(vigation and) r(anging). Figure 9.1 shows a mobile robot equipped with sonar sensors.

The process of finding your (or a robot's) location based on sonar is called
ECHOLOCATION *echolocation*. Echolocation works just the way the name sounds (no pun intended): sound bounces off objects and forms echoes that are used to find one's place in the environment. That's the basic principle. But before we get into the details, let's first consider some examples.

Figure 9.1 A mobile robot with sonar sensors.

The principle of echolocation comes from nature, where it is used by several species of animals. Bats are the most famous for using sophisticated echolocation; cave bats that dwell in nearly total darkness do not use vision (it would not do them much good), but rely entirely on ultrasound. They emit and detect different frequencies of ultrasound, which allows them to fly effectively in very crowded caves that have complicated structures and are packed with hundreds of other flying and hanging bats. They do all this very quickly and without any collisions. Besides flying around in the dark, bats also use echolocation to catch tiny insects and find mates. Dolphins are another species known for sophisticated echolocation. What used to be secret research is now standard in aquarium shows: blindfolded dolphins can find small fish and swim through hoops and mazes by using echolocation.

As usual, biological sensors are vastly more sophisticated than current artificial ones (also called "synthetic," because they have been "synthesized," not because they are made out of synthetics); bat and dolphin sonars are much more complex than artificial/synthetic sonars used in robotics and other applications. Still, synthetic sonars are quite useful, as you will see.

So how do they work?

TIME-OF-FLIGHT   Artificial ultrasound sensors, or sonars, are based on the *time-of-flight* principle, meaning they measure the time it takes something (in this case sound) to travel ("fly"). Sonars are active sensors consisting of an emitter and a detector. The emitter produces a chirp or ping of ultrasound frequency. The sound travels away from the source and, if it encounters a barrier, bounces off it (i.e., reflects from it), and perhaps returns to the receiver (microphone). If there is no barrier, the sound does not return; the sound wave weakens (attenuates) with distance and eventually breaks down.

If the sound does come back, the amount of time it takes for it to return can be used to calculate the distance between the emitter and the barrier that the sound encountered. Here is how it works: a timer is started when the chirp is emitted, and is stopped when the reflected sound returns. The resulting time is then multiplied by the speed of sound and divided by two. Why? Because the sound traveled to the barrier and back, and we are only trying to determine how far away the barrier is, its one-way distance.

SPEED OF SOUND   This computation is very simple, and relies only on knowing the *speed of sound*, which is a constant that varies only slightly due to ambient temperature. At room temperature, sound travels 1.12 feet per millisecond. Another way to put it is that sound takes 0.89 milliseconds to travel the distance of 1 foot. This is a useful constant to remember.

The hardware for sonar sensing most commonly used in robotics is the Polaroid Ultrasound Sensor, initially designed for instant cameras. (Instant cameras were popular before digital cameras were invented, since they provided instant photos; otherwise people had to wait for film to be developed, which took at least a day, unless they a personal film development lab.) The physical sensor is a round transducer, approximately 1 inch in diameter, that

TRANSDUCER   emits the chirp/ping and receives the sound (echo) that comes back. A *transducer* is a device that transforms one form of energy into another. In the case of the Polaroid (or other ultrasound) transducers, mechanical energy is converted into sound as the membrane of the transducer flexes to produce a ping that sends out a sound wave that is inaudible to humans. You can actually hear most robot sonars clicking but what you hear is the movement of the emitter (the membrane), not the sound being sent out.

The hardware (electronics) of ultrasound sensors involves relatively high power, because significant current is needed for emitting each ping. Importantly, the amount of current required is much larger than what computer processors use. This is just one of many practical examples showing why it is a good idea to separate the power electronics of a robot's sensing and actuation mechanisms from those of its controller processor. Otherwise, the robot's brain might have to literally slow down in order for the body to sense or move.

The Polaroid ultrasound sensor emits sound that spreads from a 30-degree sound cone in all directions, and at about 32 feet attenuate to a point that they do not return to the receiver, giving the sensor a 32-foot range. The range of an ultrasound sensor is determined by the signal strength of the emitter, which is designed based on the intended use of the sensor. For robots (and for instant cameras, as it happens), the range of 32 feet is typically sufficient, especially for indoor environments. Some other uses of sonar require quite a bit less or more range, as you will see in the next section.

9.1.1　Sonar Before and Beyond Robotics

Ultrasound is used in a variety of applications besides (and before) robotics, from checking on babies in utero (inside the mother's womb) to detecting objects and attackers in submarines. When sonar is used to look into people's bodies, the result is called a sonogram, echogram, or ultrasonogram, coming from the Greek *gram* meaning "letter" and referring to writing and drawing. Sound travels well through air and water, and since the human body consists largely of water (over 90 percent by weight), ultrasound is a good technology for seeing what's going on inside. One of the most common uses of sonograms is for range (distance) measurements, just like its use in robotics. However, the Polaroid and other sonars used in robotics operate at about 50 KHz, while the medical ones operate in the higher frequency range, around

HERTZ (HZ)　3.5 to 7 MHz. Just in case you did not know or you forgot, a *Hertz (Hz)* is a unit of frequency. One Hertz means once per second, a Kilohertz (KHz) is 1000 Hz and a Megahertz (MHz) is 1,000,000 Hz.

Simply sensing the distances between the emitter and the environment may be sufficient for many robotics applications, but in medical sonar imaging, much more complex postprocessing is involved in order to create a composite image of the body part. This image is not static, but is updated in real time, allowing for what looks like real-time video of, for example, the beating heart of a baby in utero.

Since sound travels well through water, while vision is almost useless in the underwater environment, it is no surprise that sonar is the favored sensor for underwater navigation, specifically for helping submarines detect and avoid any unexpected obstacles, such as other submarines. You have no doubt seen images of large, looming, menacing, and constantly ping-ing submarines in the movies. Sonars used in submarines have long ranges, through the use of stronger signal intensity and narrower cones. As in depth sounders, these sonars send an intense beam of sound into the ocean (or any body of water), and wait for it to return, in order to see how deep the water is or, in other words, how far the nearest object/surface is. As you can imagine, these sonars need to reach much farther than 32 feet; thus their use of a narrower cone and stronger signal intensity. While such uses of ultrasound are inaudible and very useful to people, they are audible and, it turns out, dangerous to marine animals, such as whales and dolphins. High-strength and long-range ultrasound emissions have been shown to confuse whales and cause them to beach and die. The exact causes of this behavior are not yet understood, but the power of ultrasound should not be underestimated. When properly controlled and directed, it can be used to break up objects, such as kidney stones. Between the body and the ocean, there are other, more mundane uses of sonar. They include automated tape measures, height measures, and burglar alarms.

The principle of time-of-flight underlies all uses of sonar as a ranging and imaging device. In almost all applications, multiple sensor units are employed for increased coverage and accuracy. Most robots using sonars are equipped with several, usually a full ring, covering a cross-section of the robot's body.

Here is an easy question: what is the smallest number of standard Polaroid sonar sensors needed to cover the cross-section of a robot?

As we will see later, a dozen sonars cover the circle around a robot (unless the robot has a very wide berth, in which case more are used). Whether a dozen or more are used, they cannot all be pinged/emitted at the same time. Can you guess why?

Sonar sensors, Polaroid or others, are inexpensive and easy to incorporate into robot hardware. If only sonars always returned accurate distance readings! But they do not, since things are never that simple in the physical world. Sonar data can be tricky, and here is the reason why.

9.1.2 Specular Reflection

As we saw, sonar sensing is based on the emitted sound wave reflecting from surfaces and returning to the receiver. But the sound wave does not necessarily bounce off the nearest surface and come right back, as we might hope it would. Instead, the direction of reflection depends on several factors, includ-

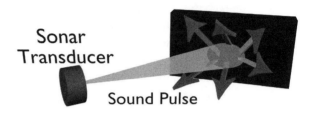

Figure 9.2   Specular reflection of an ultrasound signal.

ing the properties of the surface (how smooth it is) and the incident angle of the sound beam and the surface (how sharp it is).

A major disadvantage of ultrasound sensing is its susceptibility to specular

SPECULAR REFLECTION   reflection. *Specular reflection* is the reflection from the outer surface of the object; this means the sound wave that travels from the emitter bounces off multiple surfaces in the environment before returning to the detector. This is likely if the surface it encounters is smooth and if the angle between the beam and the surface is small. The smaller the angle, the higher the probability that the sound will merely graze the surface and bounce off, thus not returning to the emitter but moving on to other surfaces (and potentially even more grazing bounces) before returning to the detector, if it returns at all. This bouncing sound generates a false faraway reading, one that is much longer than the straight-line distance between the robot (its sonar sensor) and the surface. The smoother the surface is, the more likely the sound is to bounce off. In contrast, rough surfaces produce more irregular reflections, which are more likely to return to the emitter. Think of it this way: as the sound hits a rough surface, it scatters, bouncing back at various angles relative to the various facets and grooves and features on the surface. At least some of the reflections are likely to go back to the emitter, and thus provide a rather accurate distance measure. In contrast, as the sound hits a uniformly smooth surface (a specular one), it may graze or bounce off it uniformly in a direction away from the detector. In figure 9.2 you can see an illustration of specular reflection. Specularity is a property of light as well as of sound, which adds to the challenges of machine vision; we'll worry about that in the next section.

In the worst case of specular reflection, the sound bounces around the environment and does not return to the detector, thus fooling the sensor into detecting no object/barrier at all, or one that is far away instead of nearby. This could happen in a room full of mirrors, as at a carnival, or full of glass cases, as at a museum. A crude but effective way to combat specular reflection is to alter the environment in which the sonar-based robot has to navigate by making the surfaces less reflective. How can we do that? Well, we can rub all smooth surfaces with sandpaper, or we can use rough wallpaper, or we can put little wood slats in the walls, basically anything to introduce features on the surfaces. Fortunately, since sonar beams are relatively focused, especially at short ranges, only the surfaces that the cone of the sonars is most likely to sense need be altered, not entire walls. For example, in research labs, roboticists have lined experimental areas with bands of corrugated cardboard, because its ridges have much better sonar reflectance properties than the smooth walls. In general, altering the environment to suit the robot is not a great idea and it is often not even possible (such as under the ocean or in space).

How else can we get around the specular reflection problem?

One solution is to use phased arrays of sensors to gain more accuracy. The basic idea is to use multiple sensors covering the same physical area, but activated out of phase. This is exactly what is used in automatic tape measure devices: these contraptions, when pointed at an object, give the distance to that object, by using multiple carefully arranged and timed sonar sensors. You can think of this as a hardware-level solution to the problem.

Can you think of some software/processing-level computational solutions?

Remember that the problem is that long sonar readings can be very inaccurate, as they may result from false reflected readings from nearby objects rather than accurate faraway readings. We can use this fact to embed some intelligence into the robot, in order to make it accept short readings but do more processing on the long ones. One idea is to keep a history of past readings, and see if they get longer or shorter over time in a reasonable, continuous way. If they do, the robot can trust them, and if not, the robot assumes they are due to specular effects. This approach is effective in some environments but not in all, and is especially challenging in unknown structures. After all, the environment may have *discontinuities*, sudden and large changes in its features. Those cannot be anticipated by the robot, so it is left having to trust its sensor readings, however unpredicatable they may be.

DISCONTINUITY

Besides postprocessing, the robot can also use action-oriented perception, which we discussed in Chapter 7. Whenever it receives an unexpected long sonar reading that seems strange/unlikely, it can turn and/or move so as to change the angle between its sonar sensor and the environment, and then take another reading, or many more readings, to maximize accuracy. This is a good example of a general principle:

> *Using action to improve sensory information is a powerful method of dealing with uncertainty in robotics.*

Ultrasound sensors have been successfully used for very sophisticated robotics applications, including mapping complex outdoor terrain and indoor structures. Sonars remain a very popular, affordable ranging sensor choice in mobile robotics.

9.2   Laser Sensing

Sonars would be great sensors if they were not so susceptible to specular reflection. Fortunately, there is a sensor that largely avoids the problem, but at some cost and trade-offs: the laser.

LASER

Lasers emit highly amplified and coherent radiation at one or more frequencies. The radiation may be in the visible spectrum or not, depending on the application. For example, when laser sensors are used as burglar detectors, they are typically not visible. In the movies, we are often shown visible laser grids, but in reality making them visible makes the burglar's job easier By the way, what principle are such sensors based on? They are break beam sensors: when a burglar (or a roaming cat, say) breaks the laser beam, the alarm goes off. When laser sensors are used for range measurements, they also are typically not visible, since it is usually not desirable to have a visible and distracting beam of light scanning the environment as the robot moves around.

Laser range sensors can be used through the time-of-flight principle, just like sonars. You can immediately guess that they are much faster, since the

Figure 9.3 A mobile robot sporting a laser sensor. The laser is the thing that resembles a coffee maker and is labeled with the name of the manufacturer, SICK.

speed of light is quite a bit greater than the speed of sound. This actually causes a bit of a problem when lasers are used for measuring short distances: the light travels so fast that it comes back more quickly than it can be measured. The time intervals for short distances are on the order of nanoseconds and can't be measured with currently available electronics. As an alternative, phase-shift measurements, rather than time-of-flight, are used to compute the distance.

Robots that use lasers for indoor navigation and mapping, such as the one shown in figure 9.3, typically operate in relatively short-range environments by laser standards. Therefore, lasers used on mobile robots use phase-shift rather than time-of-flight. In both cases the processing is typically performed within the laser sensor itself, which is equipped with electronics, and so the sensor package produces clean range measurements for the robot to use.

Lasers are different from sonars in many other ways stemming from the differences in the physical properties of sound and light. Lasers involve higher-power electronics, which means they are larger and more expensive. They are also much (much, much) more accurate. For example, a popular laser sensor, the SICK LMS200 scanning laser rangefinder, has a range of 8m and a 180-degree field of view. The range and bearing of objects can be determined to within 5mm and 0.5 degrees. The laser can also be used in a long-range mode (up to 80m), which results in a reduction in accuracy of only about 10cm.

Another key distinction is that the emitted laser light is projected in a beam rather than a cone; the spot is small, about 3mm in diameter. Because lasers use light, they can take many more measurements than sonar can, thereby

RESOLUTION providing data with a higher resolution. *Resolution* refers to the process of separating or breaking something into its constituent parts. When something has high resolution, it means it has many parts. The more parts there are to the whole, the more information there is. That's why "high res" is a good thing.

So what is the resolution of the SICK sensor? Well, it makes 361 readings over a 180-degree arc at a rate of 10Hz. The true rate is much higher (again, because of the speed of light and the speed of today's electronics), but the resulting rate is imposed by the serial link for getting the data out. Ironically, the serial link is the bottleneck, but it's good enough, since a real robot does not need laser data at any rate higher than what this sensor provides via the port; it could not physically react any faster anyway.

Laser sensors sound like a dream come true, don't they? They have high resolution and high accuracy, and do not suffer nearly as much from specular effects. Of course they are not totally immune to specularities, since they are light-based and light is a wave that reflects, but specularities are not much of a problem due to the resolution of the sensor, especially compared with ultrasound sensing. So what is the downside?

Well, first of all, they are large, about the size of an electric coffee maker (and sort of look like one, too); it takes some space to house all those high-power electronics. Next, and most impressively, laser sensors are very expensive. A single SICK laser costs two orders of magnitude more than a Polaroid sonar sensor. Fortunately, price is bound to come down over time, and packaging is bound to become smaller. However, the high resolution of the laser range sensor has its downside. While the narrow beam is ideal for detecting the distance of a particular point, to cover an area, the laser needs to sweep or scan. The planar (2D) SICK laser mentioned above scans horizontally over a 180-degree range, providing a highly accurate slice of distance readings. However, if the robot needs to know about distances outside of the plane, more lasers and/or more readings are needed. This is quite doable, since lasers can scan quickly, but it does take additional sensing and processing time. 3D laser scanners also exist, but are even larger and more expensive. They are ideal, however, for accurately mapping out distances to objects, and thus the space around the sensor (or robot using the sensor).

In mobile robotics, simple sensors are usually most popular, and lasers, even planar ones, are not sufficiently affordable or portable for some applications. Can you guess which ones? For example, any that include small robots, or robots interacting with children or people who might look in the direction of the laser, which is too high-power to be considered completely safe.

Let's go back to the laser grid for detecting intruders and burglars in movies. Projecting a visible grid of laser light on the environment is part of another approach to sensing. The distortions of the grid represent the shapes of objects in the environment. But to detect the pattern and its distortions, we need another sensor: a camera. That brings us to the most complex and versatile sensory modality of all: vision.

9.3 Visual Sensing

Seeing requires a visual sensor, something akin to biological eyes. Cameras are the closest thing to natural eyes that we have available among synthetic sensors. Needless to say, any/all biological eyes are more complex than any cameras we have today. But to be fair, seeing is not done by eyes only, but largely by the brain. Eyes provide the information about the incoming light patterns, and the brain processes that information in complex ways to answer questions such as "Where are my car keys?" and make decisions such as "Stop right now; there they are, between the sofa cushions!"

Although cameras and computers are a far cry from, and much less complex than, biological eyes and brains, you will soon see that the information provided by cameras is not simple in the least, and that the associated processing is some of the most complex in robotics. The research field that deals MACHINE VISION with vision in machines, including robots, is called, appropriately, *machine vision*. Robots have particular perceptual needs related to their tasks and environments, and so some parts of machine vision research are relevant and useful for robotics, while others have proven not to be, even if they are very useful for other applications. Therefore, machine vision and robotics are two separate fields of research with some overlap of interests, problems, and uses.

Traditionally, machine vision has concerned itself with answering the questions "What is that?", "Who is that?" and "Where is that?" To answer these,

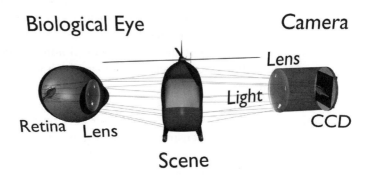

Figure 9.4 Components of biological and synthetic vision systems.

the approach has been to reconstruct what the world was like when the camera took its picture, in order to understand the picture, so that the question can be answered. We already talked about vision as reconstruction in Chapter 7.

After decades of research in machine vision, we know that visual reconstruction is an extremely difficult problem. Fortunately for robotics, it is not the problem that robots really need to solve. Instead, robots are typically concerned with acting so as to achieve their goals or, put more simply, with doing the right thing. Instead of answering the machine vision questions above, they need to answer action-related questions of the type: "Should I keep going or turn or stop?", "Should I grab or let go?", "Where do I turn?", and so on. It is usually not necessary to reconstruct the world in order to answer those questions. To begin to understand how robots use vision, let's first see (so to speak) how cameras, the basic visual sensors, do their work.

9.3.1 Cameras

BIOMIMETIC Cameras are *biomimetic* meaning they imitate biology, in that they work somewhat the way eyes do. But as usual, synthetic and natural vision sensors are quite different. Figure 9.4 shows a simple comparison, marking the key components of the vision system.

Here are those components. Light, scattered from objects in the environment (which are collectively called the *scene*), goes through an opening (the *iris*, which is in the simplest case a pinhole, but usually is a lens) and hits the image plane. The *image plane* corresponds to the retina of the biological eye, which is attached to numerous light-sensitive (*photosensitive*) elements called rods and cones. These in turn are attached to nerves that perform *early vision*, the first stages of visual image processing, and then pass information on to other parts of the brain to perform *high-level vision* processing, everything else that is done with the visual input. As we have mentioned before, a very large portion of the human (and other animal) brain activity is dedicated to visual processing, so this is a highly complex endeavor. Instead of rods and cones, film cameras use silver halides on photographic film, and digital cameras use silicon circuits in charge-coupled devices (CCD). In all cases, some information about the incoming light (e.g., its intensity, color) is detected by the photosensitive elements on the image plane.

In machine vision, the computer needs to make sense out of the information on the image plane. If the camera is very simple, and uses a tiny pinhole, then some computation is required to determine the projection of the objects from the environment onto the image plane (note that they will be inverted). If a *lens* is involved (as in vertebrate eyes and real cameras), then more light can get in, but at the price of being focused; only objects a particular range of distances from the lens will be in *focus*. This range of distances is called the camera's *depth of field*.

The image plane is usually subdivided into equal parts called *pixels*, typically arranged in a rectangular grid. As we saw in Chapter 7, in a typical camera there are 512×512 pixels on the image plane. For comparison, there are 120×10^6 rods and 6×10^6 cones in the human eye.

The projection of the scene on the image plane is called, not surprisingly, the *image*. The brightness of each pixel in the image is proportional to the amount of light that was reflected into the camera by the part of the object or surface that projects to that pixel, called the *surface patch*.

Since you already know that reflectance properties vary, you can guess that the particular reflectance properties of the surface patch, along with the number and positions of light sources in the environment, and the amount of light reflected from other objects in the scene onto the surface patch all have a strong impact on what the pixel brightness value ends up being. All those influences that affect the brightness of the patch can be lumped into

SCENE
IRIS
IMAGE PLANE
PHOTOSENSITIVE
EARLY VISION

HIGH-LEVEL VISION

LENS

FOCUS
DEPTH OF FIELD
PIXEL

IMAGE

SURFACE PATCH

two kinds of reflections: specular (off the surface, as we saw before) and
DIFFUSE REFLECTION   diffuse. *Diffuse reflection* consists of the light that penetrates into the object, is
absorbed, and then comes back out. To correctly model light reflection and
reconstruct the scene, all these properties are necessary. No wonder visual
reconstruction is hard to do. It's a good thing that robots usually don't need
to do it.

We need to step back for just a second here and remember that the camera, just like the human eye, observes the world continually. This means
TIME SERIES   it captures a video, a series of images over time. Processing any *time series*
information over time is pretty complicated. In the case of machine vision,
FRAME   each individual snapshot in time is called a *frame*, and getting frames out of
a time series is not simple. In fact, it involves specialized hardware, called
FRAME GRABBER   a *frame grabber*, a device that captures a single frame from a camera's analog
DIGITAL IMAGE   video signal and stores it as a *digital image*. Now we are ready to proceed
IMAGE PROCESSING   with the next step of visual processing, called *image processing*.

9.3.2   Edge Detection

EDGE DETECTION   The typical first step (early vision) in image processing is to perform *edge
detection*, to find all the edges in the image.

How do we recognize edges? What are edges, really?

EDGE   In machine vision, an *edge* is defined as a curve in the image plane across
which there is a significant change in the brightness. More intuitively, finding
edges is about finding sharp changes in pixel brightness. Finding changes
mathematically is done by taking derivatives. (Calculus has its uses.) A
simple approach to finding edges, then, is to differentiate the image and look
for areas where the magnitude of the derivative is large, which indicates that
the difference in the local brightness values is also large, likely due to an
edge. This does find edges, but it also finds all sorts of other things that
produce large changes, such as shadows and noise. Since it is impossible
to distinguish "real" edges from those resulting from shadows simply by
looking at pixel brightness/intensity in the image, some other method has to
be used to do better.

What about noise – how do we deal with it?

Unlike shadows, noise produces sudden and spurious intensity changes
that do not have any meaningful structure. This is actually a good thing,

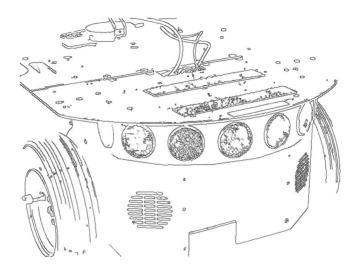

Figure 9.5 A camera image that has been processed through edge detection and then segmentation.

since noise appears as peaks in the intensities, and those peaks can be taken away by a process called "smoothing."

How is smoothing done automatically?

SMOOTHING

Again, math comes to the rescue. To perform *smoothing*, we apply a mathematical procedure called convolution, which finds and eliminates the isolated peaks. *Convolution* applies a filter to the image; this is called *convolving* the image. This type of mathematical filter is really the same, in principle, as a physical filter, in that the idea is to filter out the unwanted things (in this case the spurious peaks that come from visual noise) and let through the good stuff (in this case the real edges). The process of finding real edges involves convolving the image with many filters with different orientations. Think back to the previous chapter, when we talked about polarized light filters. Those were physical filters, and here we are talking about mathematical filters, but both have the same function: to separate out a particular part of the signal. In the case of the polarized filters, we were looking for a particular frequency of light; in the case of edge detection filters, we are looking for intensities with particular orientations.

CONVOLUTION

Edge detection used to be a very popular difficult problem in machine vision, and many algorithms were written for it, tested, and published. Eventually, researchers developed the best algorithms possible for the task, so now edge detection is not considered an interesting research problem, but it is still a very real practical problem in machine vision. Whenever one has to do edge detection, there is an "off-the-shelf" algorithm to use, and in some cases specialized hardware, such as edge detection processors, can be used to speed up the visual processing.

SEGMENTATION

Once we have edges, the next thing to do is try to find objects among all those edges. *Segmentation* is the process of dividing or organizing the image into parts that correspond to continuous objects. Figure 9.5 shows an image that has been processed by edge detection and segmentation.

But how do we know which lines correspond to which objects? And what makes an object?

The next few sections describe several cues and approaches we can use to detect objects.

9.3.3 Model-Based Vision

Suppose that your robot has a bunch of line drawings of chairs in its memory. Whenever it sees an object in the environment, it performs edge detection, which produces something like a very bad line drawing, and compares the outcome with those stored drawings to see if any of them match what it saw in the environment, which would indicate it saw a chair.

MODEL-BASED VISION

Those stored drawings are called *models* and the process is called *model-based vision*. It is a part of a philosophy about how the brain may recognize familiar objects, or how we might get robots to do it effectively. Model-based vision uses models of objects, and some prior information or knowledge about those objects, represented and stored in a way that can be used for comparison and recognition.

Models can be stored in a variety of forms; line drawings are just one form. Even though 2D line drawings are relatively simple and intuitive, using model matching for recognizing them is still a complex process. Here is why: it is not enough just to compare what the robot sees (even after edge detection) with the stored model. The robot may be looking at the object from any angle and any distance. Therefore, to compare it effectively with the model, it has to properly scale (change the size of) the model, and rotate the model to try different orientations. Also, since any edge in the image may correspond to any edge in the model, all those combinations have to be considered and evaluated. Finally, since the robot does not know what it is looking at, it needs to consider all models it has stored in its memory, unless it can somehow cleverly eliminate some of them as unlikely. All of this is computationally intensive, taking plenty of memory (to store the models) and processor power (to do scaling, rotation, and comparisons).

Models can vary from simple 2D line drawings to weirdly processed, mathematically distorted images that combine all the various views of the object to be recognized in a mathematical way. For example, some very successful face recognition systems use only a few views of the person's face, and then do some interesting math to produce a model that can then recognize that person from many more points of view. Face recognition is a very popular problem in machine vision, and model-based approaches seem very well suited for it, since faces do have repeatable features, such as two eyes, a nose, and a mouth, with relatively constant ratios of distances in between those features (for most people, at least). Nevertheless, we are still far from being able to reliably and efficiently recognize a particular "face in the crowd," whether it be your mother who has come to pick you up at the airport or a known criminal trying to flee the country.

Face recognition is one of the important things that your brain performs very effectively, and is in fact fine-tuned through evolution to do very well. This is so because humans are such *social animals*, for whom it is very important to know who is who in order to establish and maintain social order. Imagine if you could not recognize faces, what would your life be like? There is a neurological disorder that produces that very deficit in some people; it PROSOPAGNOSIA is called *prosopagnosia*, from the Greek *prosop* for "face" and *agnosia* for "not knowing." It is rare and, so far, incurable.

Face recognition is likely to be useful in robotics, especially for robots that are going to interact with people as their companions, helpers, nurses, coaches, teachers, or pets. But there are many faces to learn and recognize, so this will remain an interesting research challenge, not just for machine vision HUMAN-ROBOT but also for the area of robotics that deals with *human-robot interaction* and INTERACTION which we will discuss in Chapter 22.

9.3.4 Motion Vision

Visual systems are often, though not always, attached to things that move (such as people and robots, for example). The movement of the body and MOTION VISION the camera on it makes vision processing more challenging and *motion vision* is a set of machine vision approaches that uses motion to facilitate visual processing.

If the vision system is trying to recognize static objects, it can take advantage of its own motion. By looking at an image at two consecutive time steps, and moving the camera in between, continuous solid objects (at least those that obey physical laws we know about) will move as one, and their brightness properties will be unchanged. Therefore, if we subtract two consecutive images from one another, what we get is the "movement" between the two, while the objects stay the same. Notice that this depends on knowing exactly how we moved the camera relative to the scene (the direction and distance of the movement), and on not having anything else moving in the scene.

As we mentioned in Chapter 7, by using active perception a robot can use movement to get a better view of something. However, moving about to go places and moving to see better are not necessarily the same, and may be in conflict in some cases. Therefore, a mobile robot using vision has to make some clever decisions about how it moves, and has to subtract its own movement from the visual image in order to see what it can see.

If other objects also move in the environment, such as other robots and people, the vision problem becomes much harder. We will not go into any more details about it here, but you can pursue it in the readings at the end of this chapter.

9.3.5   Stereo Vision

BINOCULAR VISION
STEREO VISION

Everything we have discussed so far about vision has assumed a single camera. Yet in nature, creatures have two eyes, giving them *binocular vision*. The main advantage of having two eyes is the ability to see *in stereo*. *Stereo vision*, formally called *binocular stereopsis*, is the ability to use the combined points of view from the two eyes or camers to reconstruct three-dimensional solid objects and to perceive depth. The term *stereo* comes from the Greek *stereos* meaning "solid," and so it applies to any process of reconstructing the solid from multiple signals.

In stereo vision, just as in motion vision (but without having to actually move), we get two images, which we can subtract from one another, as long as we know how the two cameras or eyes are positioned relative to each other. The human brain "knows" how the eyes are positioned, and similarly we, as robot designers, have control over how the cameras on the robot are positioned as well and can reconstruct depth from the two images. So if you can afford two cameras, you can get depth perception and reconstruct solid objects.

This is the way 3D glasses, which make images in movies look solid, work. In normal movies, the images come from a single projector and both of your eyes see the same image. In 3D movies, however, there are two different images from two different projectors. That's why when you try to watch a 3D movie without the special glasses, it looks blurry. The two images do not come together, on the screen or in your brain. But when you put on the glasses, the two come together in your brain and look 3D. How? The special glasses let only one of the projected images into each of your eyes, and your brain fuses the images from the two eyes, as it does for everything you look at. This does not work with normal movies because the images in your two eyes are the same, and when they are fused, they still look the same. But the 3D movie images are different, by clever design, and when brought together, they look better.

You might wonder how it is that those simple glasses with colored foil lenses manage to make just one of the projected images go into each of your eyes. That's simple: one of the projected images is blue (or green) and one is red, and the glasses let in one color each. Pretty easy and neat, isn't it? The first 3D movie using this method was made in 1922, and you can still see the same technology used for TV, movies, and books. Today there are more sophisticated ways of achieving the same result, with better color and sharpness. These involve the use of something else you already know about: polarized light, which we learned about in Chapter 8. The movies are shown from two projectors that use different polarization, and the 3D glasses use polarizing filters instead of the simpler color filters.

The ability to perceive in 3D using stereo is fundamental to realistic human/animal vision, and so it is involved in a variety of applications from video games to teleoperated surgery. If you lose the use of one of your eyes, you will lose the ability to see depth and 3D objects. To see how important depth perception is, try catching a ball with one eye closed. It's hard, but not impossible. That is because your brain compensates for the loss of depth for a little while. If you put a patch over one eye for several hours or longer, you will start to trip and fall and reach incorrectly toward objects. Fortunately, once you take the patch off, your brain will readjust to seeing in 3D. But if you lose an eye permanently, it won't. So eyes and cameras are to be treated with care.

9.3.6   Texture, Shading, Contours

What other properties of the image can we find and use to help in object detection?

Consider texture. Sandpaper looks quite a bit different from fur, which looks quite a bit different from feathers, which look quite a bit different from a smooth mirror, and so on, because all reflect the light in very different ways. Surface patches that have uniform texture have consistent and almost identical brightness in the image, so we can assume they come from the same object. By extracting and combining patches with uniform and consistent texture, we can get a hint about what parts of the image may belong to the same object in the scene.

Somewhat similarly, shading, contours, and object shape can also be used to help simplify vision. In fact, anything at all that can be reliably extracted from a visual image has been used to help deal with the object recognition problem. This is true not only for machines but also (and first) for biological vision systems, so let's consider those now.

9.3.7   Biological Vision

The brain does an excellent job of quickly extracting the information we need from the scene. We use *model-based vision* to recognize objects and people we know. Without it, we find it hard to recognize entirely unexpected objects or novel ones, or to orient ourselves, as in the typical example of waking up and not knowing where you are. Biological model-based vision is of course different from machine vision, and it is still poorly understood, but it works remarkably well, as you can tell when you effortlessly recognize a face in the crowd or find a lost object in a pile of other stuff.

VESTIBULAR OCULAR REFLEX (VOR)
We use *motion vision* in a variety of ways in order to better understand the world around us, as well as to be able to move around while looking and not having it all result in a big blur. The latter is done through the *vestibular ocular reflex* (VOR, in which your eyes stay fixed even though your head is moving, in order to stabilize the image. (Go ahead and try it, move your head around as you read the rest of this paragraph.) There has been a great deal of research on VOR in neuroscience and machine vision and robotics, since that ability would be very useful for robots, but it is not quite simple to implement.

We are "hard-wired" to be sensitive to movement at the periphery of our field of view, as well as to looming objects, because they both indicate potential danger. Like all carnivores, we have *stereo vision*, because it helps to find and track prey. In contrast, herbivores have eyes on the sides of their heads, pointing in different directions, which are effective at scanning for (carnivorous) predators, but whose images do not overlap or get fused together in the way carnivore stereo vision works.

We are very good at recognizing shadows, textures, contours, and various other shapes. In a famous experiment performed by a scientist named Johansson in the 1970s, a few dots of light were attached to people's clothes and the people were videotaped as they moved in the dark, so only the movement of the dots was visible. Any person watching the dots could immediately tell that they were attached to moving humans, even if only very few light dots were used. This tells us that our brains are wired to recognize human motion, even with very little information. Incidentally, this depends on seeing the dots/people from the side. If the view is from the top, we are not able to readily recognize the activity. This is because our brains are not wired to observe and recognize from a top-down view. Birds' brains probably are, so somebody should do that experiment.

We have gone from machine vision to human vision to bird vision, so let's return to robot vision now.

9.3.8 Vision for Robots

Robot vision has more stringent requirements than some other applications of machine vision, and only slightly less demanding requirements than biological vision. Robot vision needs to inform the robot about important things: if it's about to fall down the stairs, if there is a human around to help/track/avoid, if it has finished its job, and so on. Since vision processing can be a very complex problem, responding quickly to the demands of the real world based on vision information is very difficult. It is not only impractical to try to perform all the above steps of image processing before the robot gets run over by a truck or falls down the stairs, but fortunately it may be unnecessary. There are good ways of simplifying the problem. Here are some of them:

1. Use color; look for specifically and uniquely colored objects, and recognize them that way (such as stop signs, human skin, etc.).

BLOB TRACKING   2. Use the combination of color and movement; this is called color *blob track-*

SALIENT

ing and is quite popular in mobile robotics. By marking important objects (people, other robots, doors, etc.) with *salient* (meaning "noticeable," "attention-getting"), or at least recognizable colors, and using movement to track them, robots can effectively get their work done without having to actually recognize objects.

3. Use a small image plane; instead of a full 512×512 pixel array, we can reduce our view to much less, for example, just a line (as is used in linear CCD cameras). Of course there is much less information in such a reduced image, but if we are clever and know what to expect, we can process what we see quickly and usefully.

4. Combine other, simpler and faster sensors with vision. For example, IR cameras isolate people by using body temperature, after which vision can be applied to try to recognize the person. Grippers allow us to touch and move objects to help the camera get a better view. Possibilities are endless.

5. Use knowledge about the environment; if the robot is driving on a road marked with white or yellow lines, it can look specifically for those lines in the appropriate places in the image. This greatly simplifies following a road, and is in fact how the first, and still some of the fastest, robot road and highway driving is done.

Those and many other clever techniques are used in robot vision to make it possible for robots to see what they need to see quickly enough for doing their task.

Consider the task of autonomous or at least semiautonomous driving. This robotics problem is gaining popularity with the auto industry, as a potential means of decreasing the number of auto accidents. Automakers would be glad to have cars that can make sure the driver does not swerve off the road or into oncoming traffic. But in that task environment, everything is moving very quickly, and there is no time for slow vision processing. This is in fact a very exciting area of machine vision and robotics research. In 2006, several robot cars (regular cars and vans with robotic control of the steering) managed to drive completely autonomously from Los Angeles to Las Vegas, a very long trip. The cars used vision and laser sensing to obtain the information needed for performing the task. The next challenge being pursued is to do something similar in urban environments.

Complex sensors imply complex processing, so they should be used selectively, for tasks where they are required or truly useful. As we have said before (and will say again):

To design an effective robot, it is necessary to have a good match between the robot's sensors, task, and environment.

To Summarize

- Sensor complexity is based on the amount of processing the data require. Sensors may also have complex mechanisms, but that is not what we are as concerned with in robotics.

- Ultrasound (sonar) sensing uses the time-of-flight principle to measure the distance between the transducer and the nearest object(s).

- Ultrasound sensing is relatively high-power and is sensitive to specular reflections.

- Ultrasound is used not only by robots and other machines (from submarines to medical imaging equipment) but also by animals (dolphins, whales).

- Lasers are used in ways similar to sonars, but are much faster and more accurate, as well as much more expensive.

- Vision is the most complex and sophisticated sensory modality, both in biology and in robotics. It requires by far the most processing and provides by far the most useful information.

- Machine vision has traditionally concerned itself with questions of recognition such as "Who is that?" and "What is that?", while robot vision has concerned itself with questions related to action, such as "Where do I go?" and "Can I grab that?"

- Object recognition is a complex problem. Fortunately, it can often be avoided in robot vision.

- Motion vision, stereo vision, model-based vision, active vision, and other strategies are employed to simplify the vision problem.

Food for Thought

- What is the speed of sound in metric units?

- How much greater is the speed of light than the speed of sound? What does this tell you about sensors that use one or the other?

- What happens when multiple robots need to work together and all have sonar sensors? How might you deal with their sensor interference? In Chapter 20 we will learn about coordinating teams of robots.

- Besides using time-of-flight, the other way to use sonars is to employ the Doppler shift. This involves examining the shift in frequency between the sent and reflected sound waves. By examining this shift, one can very accurately estimate the velocity of an object. In medical applications, sonars are used in this way to measure blood flow, among other things. Why don't we use this in robotics?

- Since two eyes are much better than one, are three eyes much better, or even any better, than two?

Looking for More?

- Check out *Directed Sonar Sensing for Mobile Robot Navigation* and other work by Hugh Durrant-Whyte (Australian Center for Field Robotics) and John Leonard (MIT) on complex sonar processing for navigation, in the lab, outdoors, and even underwater.

- You can learn more about 3D glasses here: http://science.howstuffworks.com/3-d-glasses2.htm.

- If you to discover the nitty gritty of the mathematics behind robot vision, read *Robot Vision* by Berthold Klaus Paul Horn. A good book on machine vision to check out is *Computer Vision* by Dana Ballard and Christopher Brown. There are quite a few other books on machine/robot/computer vision out there, so do a search on the Internet and browse.

10 *Stay in Control*
Feedback Control

So far, we have talked about robot bodies, including their sensors and effectors. Now it's time to turn to robot brains, the controllers that make decisions and command the robot's actions.

From Chapter 2 you remember that control theory was one of the founding areas of robotics, and that feedback control is a key component of every real robot. In this chapter we will learn the principles of feedback control and a bit of math that is used to implement feedback controllers on any system, from a steam engine to a modern robot.

10.1 Feedback or Closed Loop Control

FEEDBACK CONTROL
SET POINT

Feedback control is a means of getting a system (a robot) to achieve and maintain a desired state, usually called the *set point*, by continuously comparing its current state with its desired state.

FEEDBACK

Feedback refers to the information that is sent back, literally "fed back," into the system's controller.

The most popular example of a control system is the thermostat. It is a challenge to describe concepts in control theory without resorting to thermostats, but let's try, since we already know that thermostats are not robots.

DESIRED STATE
GOAL STATE

The *desired state* of the system, also called the *goal state*, is where the system wants to be. Not surprisingly, the notion of goal state is quite fundamental to goal-driven systems, and so it is used both in control theory and in AI, two very different fields, as we saw in Chapter 2. In AI, goals are divided into two types: achievement and maintenance.

ACHIEVEMENT GOAL

Achievement goals are states the system tries to reach, such as a particular location, perhaps the end of a maze. Once the system is there, it is done, and need not do any more work. AI has traditionally (but not exclusively) concerned itself with achievement goals.

MAINTENANCE GOAL   *Maintenance goals,* on the other hand, require ongoing active work on the part of the system. Keeping a biped robot balanced and walking, for example, is a maintenance goal. If the robot stops, it falls over and is no longer maintaining its goal of walking while balanced. Similarly, following a wall is a maintenance goal. If the robot stops, it is no longer maintaining its goal of following a wall. Control theory has traditionally (but not exclusively) concerned itself with maintenance goals.

The goal state of a system may be related to internal or external state, or a combination of both. For example, the robot's internal goal may be to keep its battery power level in some desired range of values, and to recharge the battery whenever it starts to get too low. In contrast, the robot's external goal state may be to get to a particular destination, such as the kitchen. Some goal states are combinations of both, such as the goal state that requires the robot to keep its arm extended and balance a pole. The arm state is internal (although externally observable), and the state of the pole is external. The goal state can be arbitrarily complex and consist of a variety of requirements and constraints. Anything the robot is capable of achieving and maintaining is fair game as a goal. Even what is not doable for a robot can be used as a goal, albeit an unreachable one. The robot may just keep trying and never get there. It's good to strive.

So, if the system's current and desired states are the same, it does not need to do anything. But if they are not, which is the case most of the time, how does the system decide what to do? That is what the design of the controller is all about.

10.2   The Many Faces of Error

ERROR   The difference between the current and desired states of a system is called the *error*, and a goal of any control system is to minimize that error. Feedback control explicitly computes and tells the system the error in order to help it reach the goal. When the error is zero (or small enough), the goal state is reached. Figure 10.1 shows a typical schematic of a feedback control system.

As a real world example of error and feedback, let's consider the game that's sometimes called "hot and cold," in which you have to find or guess some hidden object, and your friends help you by saying things like "You're getting warmer, hotter, colder, freezing" and so on. (Ok, we are talking about

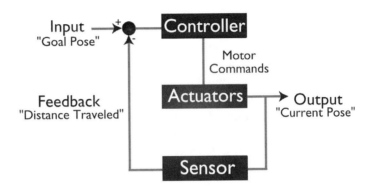

Figure 10.1   A diagram of a typical feedback controller.

temperature now, but we are still not talking about thermostats.) What they are really doing is computing the error and giving you feedback about it.

Imagine a cruel version of the same game, in which your friends tell you only "You are there, you win!" or "Nope, you are not there." In that case, what they are telling you is only if the error is zero or non-zero, if you are at the goal state or not. This is not very much information, since it does not help you figure out which way to go in order to get closer to the goal, to minimize the error.

Knowing if the error is zero or non-zero is better than knowing nothing, but not by much. In the normal version of the game, when you are told "warm" or "cool," you are being given the *direction of the error*, which allows for minimizing the error and getting closer to the goal.

DIRECTION OF ERROR

By the way, in Chapter 21 we will learn how the notion of feedback is related to the basic principles of reinforcement learning. Replace "warm" with "good robot" and "cool" with "bad robot," and you will start to get the idea of how some forms of learning work. But we have a few chapters to go before we get there.

MAGNITUDE OF ERROR

When the system knows how far off it is from the goal, it knows the *mag-*

nitude of error, the distance to the goal state. In the "hot and cold" game, the gradations of freezing, chilled, cool, warm, and so on are used to indicate the distance from (or closeness to) the goal object.

We can do better still by telling more than just the magnitude of the error. If the system is given the precise direction and magnitude of the error, it is told exactly what to do to get to the goal. In the game above, if you told the person who was guessing: "Ok, turn toward the door, go through to the kitchen, open the freezer, and look under the ice tray," it would not be much of a game. For robots, on the other hand, getting that much information still does not make control or learning trivial, but it does make it easier.

As you can see, control is made easier if the robot is given plenty of feedback information about its error and if that information is provided accurately and frequently. Let's work this out through an example of a wall-following robot.

10.3   An Example of a Feedback Control Robot

How would you write a controller for a wall-following robot using feedback control?

The first step is to consider the goal of the task. In wall-following, the goal state is a particular distance, or range of distances, from a wall. This is a maintenance goal, since wall-following involves keeping that distance over time.

Given the goal, it is simple to work out the error. In the case of wall-following, the error is the difference between the desired distance from the wall and the actual distance at any point in time. Whenever the robot is at the desired distance (or range of distances), it is in the goal state. Otherwise, it is not.

We are now ready to write the controller, but before we do that, we should consider sensors, since they will have to provide the information from which both state and error will be computed.

What sensor(s) would you use for a wall-following robot and what information would they provide?

Would they provide magnitude and direction of the error, or just magnitude, or neither? For example, a contact bump sensor would provide the least information (it's a simple sensor, after all). Such a sensor would tell the robot only whether it has hit a wall; the robot could sense the wall only through contact, not at a distance. An infra red sensor would provide information about a possible wall, but not the exact distance to it. A sonar would provide distance, as would a laser. A stereo vision system could also provide distance and even allow for reconstructing more about the wall, but that would most definitely be overkill for the task. As you can see, the sensors determine what type of feedback can be available to the robot.

Whatever sensor is used, assume that it provides the information to compute *distance-to-wall*. The robot's goal is to keep distance-to-wall at a particular value or, more realistically, in a particular range of values (let's say between 2 and 3 feet). Now we can write the robot's feedback controller in the form of standard if-then-else conditional statements used in programming, like this one:

```
If distance-to-wall is in the right range,
    then keep going.
If distance-to-wall is larger than desired,
    then turn toward the wall,
    else turn away from the wall.
```

Given the above controller algorithm, what will the robot's behavior look like? It will keep moving and switch/wiggle back and forth as it moves along. How much switching back and forth will it do? That depends on two parameters: how often the error is computed and how much of a correction (turn) is made each time.

Consider the following controller:

```
If distance-to-wall is exactly as desired,
    then keep going.
If distance-to-wall is larger than desired,
    then turn by 45 degrees toward the wall,
    else turn by 45 degrees away from the wall.
```

The above controller is not very smart. Why? To visualize it, draw a wall and a robot and follow the rules of the above controller through a few iterations to see what the robot's path looks like. It oscillates a great deal and rarely if ever reaches the desired distance before getting too close to or too far from the wall.

In general, the behavior of any simple feedback system oscillates around the desired state. (Yes, even in the case of thermostats.) Therefore, the robot oscillates around the desired distance from the wall; most of the time it is either too close or too far away.

How can we decrease this oscillation?

There are a few things we can do. The first is to compute the error often, so the robot can turn often rather than rarely. Another is to adjust the turning angle so the robot turns by small rather than large angles. Still another is to find just the right range of distances that defines the robot's goal. Deciding how often to compute the error, how large a turning angle to use, and how to define the range of distances all depend on the specific parameters of the robot system: the robot's speed of movement, the range of the sensor(s), and the rate with which new distance-to-wall is sensed and computed, called the

SAMPLING RATE   *sampling rate.*

> *The calibration of the control parameters is a necessary, very important, and time-consuming part of designing robot controllers.*

Control theory provides a formal framework for how to properly use parameters in order to make controllers effective. Let's learn how.

10.4   Types of Feedback Control

The three most used types feedback control are proportional control (P), proportional derivative control (PD), and proportional integral derivative control (PID). These are commonly referred to as P, PD, and PID control. Let's learn about each of them and use our wall-following robot as an example system.

10.4.1   Proportional Control

PROPORTIONAL   The basic idea of *proportional control* is to have the system respond in propor-
CONTROL   tion to the error, using both the direction and the magnitude of the error. A proportional controller produces an output o proportional to its input i, and is formally written as:

$$o = K_p i$$

K_p is a proportionality constant, usually a constant that makes things work, and is specific to a particular control system. Such parameters are abundant in control; typically you have to figure them out by trial and error and calibration.

What would a proportional controller for our wall fallowing robot look like?

It would use distance-to-wall as a parameter to determine the angle and distance and/or speed with which the robot would turn. The larger the error, the larger the turning angle and speed and/or distance; the smaller the error, the smaller the turning angle and speed and/or distance.

GAIN

In control theory, the parameters that determine the magnitude of the system's response are called *gains*. Determining the right gains is typically very difficult, and requires trial and error, testing and calibrating the system repeatedly. In some cases, if the system is very well understood, gains can be computed mathematically, but that is rare.

PROPORTIONAL GAIN

If the value of a particular gain is proportional to that of the error, it is called *proportional gain*. As we saw in the case of our wall-following robot, incorrect gain values cause the system to undershoot or overshoot the desired state. The gain values determine whether the robot will keep oscillating or will eventually settle on the desired state.

DAMPING

Damping refers to the process of systematically decreasing oscillations. A system is properly *damped* if it does not oscillate out of control, meaning its oscillations are either completely avoided (which is very rare) or, more practically, the oscillations gradually decrease toward the desired state within a reasonable time period. Gains have to be adjusted in order to make a system properly damped. This is a tuning process that is specific to the particular control system (robot or otherwise).

ACTUATOR UNCERTAINTY

When adjusting gains, you have to keep in mind both the physical and the computational properties of the system. For example, how the motor responds to speed commands plays a key part in control, as do the backlash and friction in the gears (remember Chapter 4?), and so on. Physical properties of the robot influence the exact values of the gains, because they constrain what the system actually does in response to a command. *Actuator uncertainty* makes it impossible for a robot (or a human, for that matter) to know the exact outcome of an action ahead of time, even for a simple action such as "Go forward three feet." While actuator uncertainty keeps us from predicting the exact outcome of actions, we can use probability to estimate and make a pretty good guess, assuming we know enough about the system to set up the probabilities correctly. See the end of the chapter for more reading on probabilistic robotics.

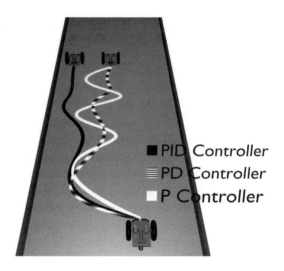

Figure 10.2 Different trajectories produced by P, PD, and PID feedback controllers.

10.4.2 Derivative Control

As we have seen, setting gains is difficult, and simply increasing the proportional gain does not remove oscillatory problems from a control system. While this may work for small gains (called low gains; gain is referred to as high or low), as the gain increases, the system's oscillations increase along with it. The basic problem has to do with the distance from the set point/desired state: *When the system is close to the desired state, it needs to be controlled differently than when it is far from it.* Otherwise, the momentum generated by the controller's response to the error, its own correction, carries the system beyond the desired state and causes oscillations. One solution to this problem is to correct the momentum as the system approaches the desired state.

First, let's remember what momentum is:

$$Momentum = mass * velocity$$

Since momentum and velocity are directly proportional (the faster you move and/or the bigger you are, the more momentum you have), we can control momentum by controlling the velocity of the system. As the system nears the desired state, we subtract an amount proportional to the velocity:

$$-(gain * velocity)$$

DERIVATIVE TERM   This is called the *derivative term*, because velocity is the derivative (the rate of change) of position. Thus, a controller that has a derivative term is called a D controller.

A derivative controller produces an output o proportional to the derivative of its input i:

$$o = K_d \frac{di}{dt}$$

K_d is a proportionality constant, as before, but this time with a different name, so don't assume you can use the same number in both equations.

The intuition behind derivative control is that the controller corrects for the momentum of the system as it approaches the desired state. Let's apply this idea to our wall-following robot. A derivative controller would slow the robot down and decrease the angle of its turning as its distance from the wall gets closer to the desired state, the optimal distance to the wall.

10.4.3   Integral Control

Yet another level of improvement can be made to a control system by intro-
INTEGRAL TERM   ducing the so-called *integral term*, or I. The idea is that the system keeps track of its own errors, in particular of the repeatable, fixed errors that are
STEADY STATE ERROR   called *steady state errors*. The system integrates (sums up) these incremental errors over time, and once they reach some predetermined threshold (once the cumulative error gets large enough), the system does something to compensate/correct.

An integral controller produces an output o proportional to the integral of its input i:

$$o = K_f \int i(t)dt$$

K_f is a proportionality constant. You get the pattern.

How do we apply integral control to our wall-following robot?

Actually, it's not too easy, because there is no obvious place for steady state error to build up in the simple controller we came up with. That's a good thing about our controller, so we can pat ourselves on the back and take another example. Consider a lawn-mowing robot which carefully covers the complete lawn by going from one side of the yard to the other, and moving over by a little bit every time to cover the next strip of grass in the yard. Now suppose that the robot has some consistent error in its turning mechanism, so whenever it tries to turn by a 90-degree angle to move over to the next strip on the grass, it actually turns by a smaller angle. This makes it fail to cover the yard completely, and the longer the robot runs, the worse its coverage of the yard gets. But if it has a way of measuring its error, even only once it gets large enough (for example, by being able to detect that it is mowing already mowed grass areas a lot), it can apply integral control to recalibrate.

Now you know about P, I, and D, the basic types of feedback control. Most real-world systems actually use combinations of those three basic types; PD and PID controllers are especially prevalent, and are most commonly used in industrial applications. Let's see what they are.

10.4.4   PD and PID Control

PD control is a combination, actually simply a sum, of proportional (P) and derivative (D) control terms:

$$o = K_p i + K_d \frac{di}{dt}$$

PD control is extremely useful and applied in most industrial plants for process control.

PID control is a combination (yes, again a sum) of proportional P, integral I, and derivative D control terms:

$$o = K_p i + K_f \int i(t)dt + K_d \frac{di}{dt}$$

Figure 10.2 shows example trajectories that would be produced by our wall-following robot if it were controlled by P, PD, and PID feedback controllers.

You can learn a great deal more about feedback control and the larger topic of control theory from the sources given at the end of this chapter. It's a large field studied in electrical engineering and mechanical engineering.

Before we move on, we should put control theory into perspective relative to robotics. As you have been reading and will read much more in the chapter to come, getting robots to do something useful requires many components. Feedback control plays a role at the low level, for example, for controlling the wheels or other continuously moving actuators. But for other aspects of robot control, in particular achieving higher-level goals (navigation, coordination, interaction, collaboration, human-robot interaction),

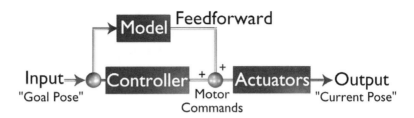

Figure 10.3   A diagram of a typical feedforward controller.

other approaches must come into play that are better suited for representing and handling those challenges. Those levels of robot control use techniques that come from the field of artificial intelligence, but that have come a long way from what AI used to be about in its early days, as we learned in Chapter 2. That's coming up, so stick around.

10.5   Feedforward or Open Loop Control

CLOSED LOOP CONTROL

Feedback control is also called *closed loop control* because it closes the loop between the input and the output, and provides the system with the error as feedback.

What is an alternative to closed loop control?

You guessed it (or you read the section title): the alternative to feedback or closed loop control is called feedforward or open loop control. As the name implies, *open loop control* or *feedforward control*, does not use sensory feedback, and state is not fed back into the system. Thus the loop between the input and output is open, and not really a loop at all. In open loop control, the system executes the command that is given, based on what was predicted, instead of looking at the state of the system and updating it as it goes along, as in feedback control. In order to decide how to act in advance, the controller determines set points or sub-goals for itself ahead of time. This requires looking ahead (or looking forward) and predicting the state of the system, which is why the approach is called feedforward. Figure 10.3 shows a schematic of a feedforward open loop controller.

OPEN LOOP CONTROL
FEEDFORWARD CONTROL

Open loop or feedforward control systems can operate effectively if they are well calibrated and their environment is predictable and does not change in a way that affects their performance. Therefore they are well suited for repetitive, state-independent tasks. As you have probably guessed, those are not very common in robotics.

To Summarize

- Feedback/closed loop control and feedforward/open loop control are important aspects of robotics.

- Feedback control aims to minimize system error, the difference between the current state and the desired state.

- The desired state or goal state is a concept used in AI as well as control theory, and can come in the form of achievement or maintenance.

- Error is a complex concept with direction and magnitude, and cumulative properties.

- Proportional, derivative, and integral control are the basic forms of feedback control, which are usually used in combination in real-world control systems (robotic and otherwise).

Food for Thought

- What happens when you have sensor error in your system? What if your sensor incorrectly tells you that the robot is far from a wall but in fact it is not? What about vice versa? How might you address these issues?

- What can you do with open loop control in your robot? When might it be useful?

Looking for More?

- The Robotics Primer Workbook exercises for this chapter are found here: http://roboticsprimer.sourceforge.net/workbook/Feedback_Control

- Here are some textbooks on control theory, if you want to learn (much) more:

  - *Signals and Systems* by Simon Haykin and Barry Van Veen, John Wiley.

- *Intelligent Control Systems, Theory and Applications* edited by Madan M. Gupta and Naresh K. Sinha.
- *Linear Control System Analysis and Design: Conventional and Modern* by J. J. D'Azzo and C. Houpis.
- *Automatic Control Systems* by B. C. Kuo.

11 *The Building Blocks of Control*
Control Architectures

The job of the controller is to provide the brains for the robot and so it can be autonomous and achieve goals. We have seen that feedback control is a very good way to write controllers for making a robot perform a single behavior, such as following a wall, avoiding obstacles, and so on. Those behaviors do not require much thinking, though. Most robots have more things to do than just follow a wall or avoid obstacles, ranging from simple survival (not running into things or running out of power) to complex task achievement (whatever it may be). Having to do multiple things at once and deciding what to do at any point in time is not simple, even for people, much less for robots. Therefore, putting together controllers that will get the robot to produce the desired overall behavior is not simple, but it is what robot control is really about.

> *So how would you put multiple feedback controllers together? What if you need more than feedback control? How would you decide what is needed, which part of the control system to use in a given situation and for how long, and what priority to assign to it?*

I hope you had no hard-and-fast answers to the above questions, because there aren't any, in general. These are some of the biggest challenges in robot control, and we are going to learn how to deal with them in this and the next few chapters.

11.1 Who Needs Control Architectures?

Just putting rules or programs together does not result in well-behaved robots, through it may be a lot of fun, as long as the robots are small and not dangerous. While there are numerous different ways in which a robot's controller can be programmed (and there are infinitely many possible robot control programs), most of them are pretty bad, ranging from completely incorrect to merely inefficient. To find a good (correct, efficient, even optimal) way to control a given robot for a given task, we need to know some guiding principles of robot control and the fundamentally different ways in which robots can be programmed. These are captured in robot control architectures.

CONTROL ARCHITECTURE

A robot *control architecture* provides guiding principles and constraints for organizing a robot's control system (its brain). It helps the designer to program the robot in a way that will produce the desired overall output behavior.

The term *architecture* is used here in the same way as in "computer architecture", where it means the set of principles for designing computers out of a collection of well-understood building blocks. Similarly, in robot architectures, there is a set of building blocks or tools at your disposal to make the job of robot control design easier. Robot architectures, like computer architectures, and of course "real" building architectures, where the term originally comes from, use specific styles, tools, constraints, and rules.

To be completely honest, you don't have to know anything about control architectures to get a robot to do something. So why should we bother with this and the next few chapters? Because there is more to robotics than getting simple robots to do simple things. If you are interested in how you can get a complex robot (or a team of robots) to do something useful and robust in a complicated environment, knowing about control architectures is necessary to help you get you there. Trial and error and intuition go only so far, and we'd like to take robotics very far indeed, far beyond that.

You might be wondering exactly what we mean by robot control, since so far we have seen that it involves hardware, signal processing, and computation. That's true in general: the robot's "brain" can be implemented with a conventional program running on a microprocessor, or it may be embedded in hardware, or it may be a combination of the two. The robot's controller does not need to be a single program on a single processor. In most robots it is far from that, since, as we have seen, there is a lot going on in a robot. Sensors, actuators, and decisions need to interact in an effective way to get the robot to do its job, but there is typically no good reason to have all those elements controlled from a single centralized program, and there are many reasons not to.

Can you think of some of those reasons?

Robustness to failure, for one thing; if the robot is controlled in a centralized fashion, the failure of that one processor makes the whole robot stop working. But before we think about how to spread out the different parts of the robot's brain, let's get back to designing robot control programs, wherever on the robot they may be running.

> *Robot control can take place in hardware and in software, but the more complex the controller is, the more likely it is to be implemented in software. Why?*

Hardware is good for fast and specialized uses, and software is good for flexible, more general programs. This means that complicated robot brains typically involve computer programs of some type or another, running on the robot in real time. The brain should be physically on the robot, but that may be just prejudice based on how biological systems do it. If wireless radio communication is reliable enough, some or all of the processing could reside off the robot. The trouble is that communication is never perfectly reliable, so it is much safer to keep your brain with you at all times (keep your head on your shoulders), for robots and people.

ALGORITHM

Brains, robotic or natural, use their programs to solve problems that stand in the way of achieving their goals and getting their jobs done. The process of solving a problem using a finite (not endless) step-by-step procedure is called an *algorithm*, and is named for the Iranian mathematician, Al-Khawarizmi (if that does not seem similar at all, it may be because we are not pronouncing his name correctly). The field of computer science devotes a great deal of research to developing and analyzing algorithms for all kinds of uses, from sorting numbers to creating, managing, and sustaining the Internet. Robotics also concerns itself with the development and analysis of algorithms (among many other things, as we will learn) for uses that are relevant to robots, such as navigation (see Chapter 19), manipulation (see Chapter 6), learning (see Chapter 21), and many others (see Chapter 22). You can think of algorithms as the structure on which computer programs are based.

11.2 Languages for Programming Robots

So robot brains are computer programs and they are written using programming languages. You may be wondering what the best robot programming language is. Don't waste your time: there is no "best" language. Robot programmers use a variety of languages, depending on what they are trying to get the robot to do, what they are used to, what hardware comes with the robot, and so on. The important thing is that robot controllers can be implemented in various languages.

Any programming language worth its beans is called "Turing universal," which means that, in theory at least, it can be used to write any program. This concept was named after Alan Turing, a famous computer scientist from England who did a great deal of foundational work in the early days of computer science, around World War II. To be Turing universal, a programming language has to have the following capabilities: sequencing (a then b then c), conditional branching (if a then b else c), and iteration (for a=1 to 10 do something). Amazingly, with just those, any language can compute everything that is computable. Proving that, and explaining what it means for something to be computable, involves some very nice formal computer science theory that you can learn about from the references at the end of the chapter.

The good news from all that theory is that programming languages are just tools; you can use them to program various things, from calculators to airline scheduling to robot behaviors, and everything in between. You can use any programming language to write any program, at least in theory. In practice, of course, you should be smart and picky, and choose the language that is well suited to the programming task. (You have no doubt noticed this repeating theme about finding suitable designs for robot sensors and bodies and controllers, and suitable languages for programming the robots, and so on. It's a basic principle of good engineering of any kind.)

If you have programmed even a little bit, you know that programming languages come in a great variety and are usually specific for particular uses. Some are good for programming Web pages, others for games, and still others for programming robots, which is what we really care about in this book. So although there is no best language for programming robots, some languages are better for it than others. As robotics is growing and maturing as a field, there are more and more specialized programming languages and tools.

Various programming languages have been used for robot control, ranging from general-purpose to specially designed ones. Some languages have been designed specifically to make the job of programming particular robot control architectures easier. Remember, architectures exist to provide guiding principles for good robot control design, so it is particularly convenient if the programming language can make following those principles easy and failing to follow them hard.

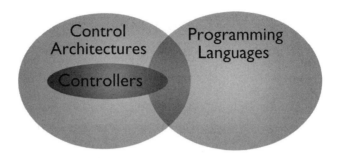

Figure 11.1 A way to visualize the relationship among control architectures, controllers, and programming languages.

Figure 11.1 shows a way in which you can think about the relationship among control architectures, robot controllers, and programming languages. They are all different, yet are needed to make a robot do what you'd like it to do.

11.3 And the Architectures are...

Regardless of which language is used to program a robot, what matters is the control architecture used to implement the controller, because not all architectures are the same. On the contrary, as you will see, architectures impose strong rules and constraints on how robot programs are structured, and the resulting control software ends up looking very different.

We have already agreed that there are numerous ways you can program a robot, and there are fewer, but still many, ways in which you can program a robot well. Conveniently, all those effective robot programs fall into one of the known types of control architectures. Better still, there are very few types of control (that we know of so far). They are:

1. Deliberative control

2. Reactive control

3. Hybrid control

4. Behavior-based control.

In the next few chapters we will study each of these architectures in detail, so you will know what's good and bad about them and which one to choose for a given robot and task.

Because robotics is a field that encompasses both engineering and science, there is a great deal of ongoing research that brings new results and discoveries. At robotics conferences, there are presentations, full of drawings of boxes and arrows, describing architectures that may be particularly well suited for some robot control problem. That's not about to go away, since there are so many things robots could do, and so much yet to be discovered. But even though people may keep discovering new architectures, all of them, and all the robot programs, fit into one of the categories listed above, even if the programmer may not realize it.

In most cases, it is impossible to tell, just by observing a robot's behavior, what control architecture it is using. That is because multiple architectures can get the job done, especially for simple robots. As we have said before, when it comes to more complex robots, the control architecture becomes very important.

Before we launch into the details of the different types of control architectures and what they are good for, let's see what are some important questions to consider that can help us decide which architecture to use. For any robot, task, and environment, many things need to be considered, including these:

- Is there a lot of sensor noise?

- Does the environment change or stay static?

- Can the robot sense all the information it needs? If not, how much can it sense?

- How quickly can the robot sense?

- How quickly can the robot act?

- Is there a lot of actuator noise?

- Does the robot need to remember the past in order to get the job done?

- Does the robot need to think into the future and predict in order to get the job done?

- Does the robot need to improve its behavior over time and be able to learn new things?

Control architectures differ fundamentally in the ways they treat the following important issues:

- Time: How fast do things happen? Do all components of the controller run at the same speed?

- Modularity: What are the components of the control system? What can talk to what?

- Representation: What does the robot know and keep in its brain?

Let's talk about each of these briefly. We will spend more time understanding them in detail in the subsequent chapters.

11.3.1 Time

TIME-SCALE   Time, usually called *time-scale*, refers to how quickly the robot has to respond to the environment compared with how quickly it can sense and think. This is a key aspect of control and therefore a major influence on the choice of what architecture should be used.

The four basic architecture types differ significantly in how they treat time.

DELIBERATIVE CONTROL
REACTIVE CONTROL   *Deliberative control* looks into the future, so it works on a long time-scale (how long depends on how far into the future it looks). In contrast, *reactive control* responds to the immediate, real-time demands of the environment without looking into the past or the future, so it works on a short time-scale.

HYBRID CONTROL   *Hybrid control* combines the long time-scale of deliberative control and the short time-scale of reactive control, with some cleverness in between. Finally,

BEHAVIOR-BASED CONTROL   *behavior-based control* works to bring the time-scales together. All of this will make more sense as we spend time getting to know each type of control and the associated architectures.

11.3.2 Modularity

MODULARITY   *Modularity* refers to the way the control system (the robot's program) is broken into pieces or components, called modules, and how those modules interact with each other to produce the robot's overall behavior. In *deliberative control*, the control system consists of multiple modules, including sensing (perception), planning, and acting, and the modules do their work in sequence, with the output of one providing the input for the next. Things happen one at a time, not at the same time, as we will see in Chapter 13. In *reactive control*, things happen at the same time, not one at a time. Multiple modules are all active in parallel and can send messages to each other in various ways, as we will learn in Chapter 14. In *hybrid control*, there are three main modules to the system: the deliberative part, the reactive part, and the part in between. The three work in parallel, at the same time, but also talk to each other, as we will learn in Chapter 15. In *behavior-based control*, there are usually more than three main modules that also work in parallel and talk to each other, but in a different way than in hybrid systems, as we will cover in Chapter 16. So as you can see, how many modules there are, what is in each module, whether the modules work sequentially or in parallel, and which modules can talk to which others are distinguishing features of control architectures.

11.3.3   Representation

Finally, *representation*. That's a tough one to summarize, so we'll give it a whole chapter, coming up next, to do it justice.

> *So how do we know which architecture to use for programming the controller of a particular robot?*

Let's learn more about the different architectures so we can answer that question. After talking about representation, in the next few chapters we will study each of the four basic architecture types: deliberative, reactive, hybrid, and behavior-based.

To Summarize

- Robot control can be done in hardware and/or in software. The more complex the robot and its task, the more software control is needed.

- Control architectures provide guiding principles for designing robot programs and control algorithms.

- There is no one best robot programming language. Robots can be programmed with a variety of languages, ranging from general-purpose to special-purpose.

- Special-purpose robot programming languages can be written to facilitate programming robots in particular control architectures.

- Robot control architectures differ substantially in how they handle time, modularity, and representation.

- The main robot control architectures are deliberative (not used), reactive, hybrid, and behavior-based.

Food for Thought

- How important is the programming language? Could it make or break a particular gadget, device, or robot?

- With the constant development of new technologies that use computation, do you think there will be increasingly more or increasingly fewer programming languages?

Looking for More?

- *Introduction to the Theory of Computation* by Michael Sipser is an excellent textbook for learning about the very abstract but very elegant topics only touched on in this chapter.

- One of the most widely used textbooks for learning about programming languages is *Structure and Interpretation of Computer Programs* by Harold Abelson and Gerald Jay Sussman. It's actually fun to read, although not necessarily easy to understand.

12 *What's in Your Head? Representation*

In many tasks and environments, the robot cannot immediately sense everything it needs to know. It is therefore sometimes useful to remember what happened in the past, or to try to predict what will happen in the future. It is also sometimes useful to store maps of the environment, images of people or places, and various other information that will be useful for getting the job done.

REPRESENTATION *Representation* is the form in which information is stored or encoded in the robot.

Representation is more than memory. In computer science and robotics,
MEMORY we think of *memory* as the storage device used to keep information. Just referring to memory does not say anything about what is stored and how it is encoded: is it in the form of numbers, names, probabilities, x,y locations, distances, colors? Representation is what encodes those important features of what is in the memory.

What is represented and how it is represented has a major impact on robot control. This is not surprising, as it is really the same as saying "What is in your brain influences what you can do." In this chapter we will learn about what representation is all about and why it has such an important role in robot control.

> *How does internal state, the information a robot system keeps around (remember Chapter 3), relate to representation? Is it the same thing? If not, what is it?*

In principle, any internal state is a form of representation. In practice, what matters is the form and function of that representation, how it is stored

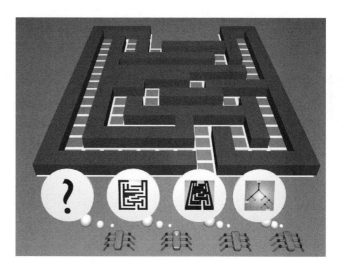

Figure 12.1   A few possible representation options for a maze-navigating robot.

and how it is used. "Internal state" usually refers to the "status" of the system itself, whereas "representation" refers to arbitrary information about the world that the robot stores.

12.1   The Many Ways to Make a Map

WORLD MODEL   Representation of the world is typically called a *world model*. A map (as used in navigation; see Chapter 19) is the most commonly used example of a world model. To give an illustration of how the representation of a particular world, its map, can vary in form, consider the problem of exploring a maze.

What can the robot store/remember to help it navigate a maze?

- The robot may remember the exact path it has taken to get to the end of the maze (e.g., "Go straight 3.4 cm, turn left 90 degrees, go straight 12 cm, turn right 90 degrees."). This remembered path is a type of map for getting through the maze. This is an odometric path.

- The robot may remember a sequence of moves it made at particular landmarks in the environment (e.g., "Left at the first junction, right at the second junction, straight at the third."). This is another way to store a path through the maze. This is a landmark-based path.

- The robot may remember what to do at each landmark in the maze (e.g., "At the green/red junction go left, at the red/blue junction go right', at the blue/orange junction go straight."). This is a landmark-based map; it is more than a path since it tells the robot what to do at each junction, no matter in what order it reaches the junction. A collection of landmarks connected with links is called a *topological map* because it describes the *topology*, the connections, among the landmarks. Topological maps are very useful. (Don't confuse them with *topographic maps*, which are something completely different, having to do with representing the elevation levels of the terrain.)

TOPOLOGICAL MAP

- The robot may remember a map of the maze by "drawing it" using exact lengths of corridors and distances between walls it sees. This is a metric map of the maze, and it is also very useful.

The above are not nearly all the ways in which the robot can construct and store a model of the maze. However, the four types of models above already show you some important differences in how representations can be used. The first model, the odometric path, is very specific and detailed, and that it is useful only if the maze never changes, no junctions become blocked or opened, and if the robot is able to keep track of distances and turns very accurately. The second approach also depends on the map not changing, but it does not require the robot to be as specific about measurements, because it relies on finding landmarks (in this case, junctions). The third model is similar to the second, but connects the various paths into a landmark-based map, a network of stored landmarks and their connections to one another. Finally, the fourth approach is the most complicated because the robot has to measure much more about the environment and store much more information. On the other hand, it is also the most generally useful, since with it the robot can now use its map to think about other possible paths in case any junctions become blocked.

Figure 12.1 illustrates a few more possible representations that could be used by a maze-following robot: no representation at all, a typical 2D map, a visual image of the environment, and a graph of the maze structure.

12.2   What Can the Robot Represent?

Maps are just one of the many things a robot may want to represent or store or model in its "brain." For example, the robot may want to remember how long its batteries last, and to remind itself to recharge them before it is too late. This is a type of a self-model. Also, the robot may want to remember that traffic is heavy at particular times of day or in particular places in the environment, and to avoid traveling at those times and in those places. Or it may store facts about other robots, such as which ones tend to be slow or fast, and so on.

There are numerous aspects of the world that a robot can represent and model, and numerous ways in which it can do it. The robot can represent information about:

- Self: stored proprioception, self-limitations, goals, intentions, plans

- Environment: navigable spaces, structures

- Objects, people, other robots: detectable things in the world

- Actions: outcomes of specific actions in the environment

- Task: what needs to be done, where, in what order, how fast, etc.

You can already see that it could take quite a bit of sensing, computation, and memory to acquire and store some types of world models. Moreover, it is not enough simply to get and store a world model; it is also necessary to keep it accurate and updated, or it becomes of little use or, worse yet, becomes misleading to the robot. Keeping a model updated takes sensing, computation, and memory.

Some models are very elaborate; they take a long time to construct and are therefore kept around for the entire lifetime of the robot's task. Detailed metric maps are such models. Other models, on the other hand, may be relatively quickly constructed, briefly used, and soon discarded. A snapshot of the robot's immediate environment showing the path out of the clutter through the nearest door is an example of such a short-term model.

12.3   Costs of Representing

In addition to constructing and updating a representation, using a representation is also not free, in terms of computation and memory cost. Consider maps again: to find a path from a particular location in the map to the goal location, the robot must plan a path. As we will learn in more detail in Chapter 19, this process involves finding all the free/navigable spaces in the map, then searching through those to find any path, or the best path, to the goal.

Because of the processing requirements involved in the construction, maintenance, and use of representation, different architectures have very different properties based on how representation is handled, as we will see in the next few chapters. Some architectures do not facilitate the use of models (or do not allow them at all, such as reactive ones), others utilize multiple types of models (hybrid ones), still others impose constraints on the time and space allowed for the models being used (behavior-based ones).

How much and what type of representation a robot should use depends on its task, its sensors, and its environment. As we will see when we learn more about particular robot control architectures, how representation is handled and how much time is involved in handling it turns out to be a crucial distinguishing feature for deciding what architecture to use for a given robot and task. So, as you would imagine, what is in the robot's head has a very big influence on what that robot can do.

To Summarize

- Representation is the form in which information is stored in the robot.

- Representation of the world around the robot is called a world model.

- Representation can take a great variety of forms and can be used in a great variety of ways by a robot.

- A robot can represent information about itself, other robots, objects, people, the environment, tasks, and actions.

- Representations require constructing and updating, which have computational and memory costs for the robot.

- Different architectures treat representation very differently, from not having any at all, to having centralized world models, to having distributed ones.

Food for Thought

- Do you think animals use internal models? What about insects?

- Why might you not want to store and use internal models?

13 *Think Hard, Act Later*
Deliberative Control

Deliberation refers to thinking hard; it is defined as "thoughtfulness in decision and action." Deliberative control grew out of early artificial intelligence (AI). As you remember from the brief history of robotics (Chapter 2), in those days, AI was one of the main influences on how robotics was done.

In AI, deliberative systems were (and sometimes still are) used for solving problems such as playing chess, where thinking hard is exactly the right thing to do. In games, and in some real-world situations, taking time to consider possible outcomes of several actions is both affordable (there is time to do it) and necessary (without strategy, things go bad). In the 1960s and 1970s, AI researchers were so fond of this type of reasoning, they theorized that the human brain works this way, and so should robot control. As you recall from Chapter 2, in the 1960s, early AI-based robots often used vision sensors, which require a great deal of processing, and so it was worth it for a robot to take the time (quite a lot of time back then, with those slow processors) to think hard about how to act given how difficult it was to make sense out of the environment. The robot Shakey, a forerunner of many AI-inspired robotics projects that followed, used the then state of the art in machine vision as input into a planner in order to decide what to do next, how and where to move.

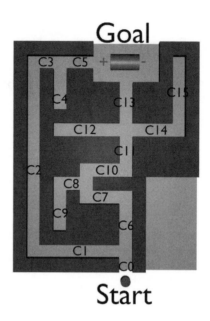

Figure 13.1 A closeup of a maze the robot (marked with a black circle) needs to navigate.

13.1   What Is Planning?

So what is planning?

PLANNING   *Planning* is the process of looking ahead at the outcomes of the possible actions, and searching for the sequence of actions that will reach the desired goal.

SEARCH   *Search* is an inherent part of planning. It involves looking through the available representation "in search of" the goal state. Sometimes searching the complete representation is necessary (which can be very slow, depending on the size of the representation), while at other times only a partial search is enough, to reach the first found solution.

For example, if a robot has a map of a maze, and knows where it is and

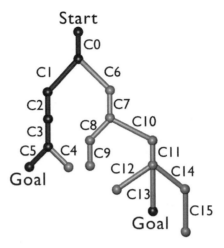

Figure 13.2   A graph of the maze, showing the two possible paths to the charger. One of the paths is shorter/more optimal than the other.

where it wants to end up (say at a recharging station at the end of the maze), it can then plan a path from its current position to the goal, such as those shown in figure 13.2. The process of searching through the maze happens in the robot's head, in the representation of the maze (such as the one shown in figure 13.1, not in the physical maze itself. The robot can search from the goal backwards, or from where it is forward. Or it can even search in both directions in parallel. That is the nice thing about using internal models or representations: you can do things that you can't do in the real world. Consider the maze in figure 13.1, where the robot is marked with a black circle, and the goal with the recharging battery. Note that at each junction in the maze, the robot has to decide how to turn.

   The process of planning involves the robot trying different turns at each junction, until a path leads it to the battery. In this particular maze, there is more than one path to the goal from where the robot is, and by searching the entire maze (in its head, of course) the robot can find both of those paths, and then choose the one it likes better. Usually the shortest path is considered the best, since the robot uses the least time and battery power to reach it. But in some cases other criteria may be used, such as which path is the safest or the least crowded. The process of improving a solution to a problem by

OPTIMIZATION   finding a better one is called *optimization*. As in the maze example above, various values or properties of the given problem can be optimized, such as

OPTIMIZATION   the distance of a path; those values are called *optimization criteria*. Usually
CRITERIA   some optimization criteria conflict (for example, the shortest path may also be the most crowded), so deciding what and how to optimize is not simple.

OPTIMIZING SEARCH   *Optimizing search* looks for multiple solutions (paths, in the case of the maze), in some cases all possible paths.

In general, in order to use search and plan a solution to a particular problem, it is necessary to represent the world as a set of states. In the maze example, the states are the corridors, the junctions, the start (where the search starts), the current state (wherever the robot happens to be at the time), and the goal (where the robot wants to end up). Then search is performed to find a path that can take the robot from the current state to the goal state. If the robot wants to find the very best, optimal path, it has to search for *all possible paths* and pick the one that is optimal based on the selected optimization criterion (or criteria, if more than one is used).

13.2   Costs of Planning

For small mazes like the one used here, planning is easy, because the state space is small. But as the number of possible states becomes large (as in chess, for example), planning becomes slow. The longer it takes to plan, the longer it takes to solve the problem. In robotics, this is particularly important since a robot must be able to avoid immediate danger, such as collisions with objects. Therefore, if path planning takes too long, the robot either has to stop and wait for planning to finish before moving on, or it may risk collisions or running into blocked paths if it forges ahead without a finished plan. This is not just a problem in robotics, of course; *whenever a large state space is involved, planning is difficult.* To deal with this fundamental problem, AI researchers have found various ways to speed things up. One popular approach is to use hierarchies of states, where first only a small number of "large," "coarse," or "abstract" states is considered; after that, more refined and detailed states are used in the parts of the state space where it really matters. The graph of the maze shown in figure 13.2 is an example of doing exactly that: it lumps all states within a corridor of the maze together and considers them as a single corridor state in the graph. There are several other clever methods

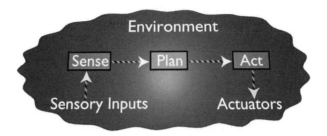

Figure 13.3   A diagram of a classical deliberative architecture, showing the sequence of sense-plan-act (SPA) components.

besides this one that can speed up search and planning. These are all optimization methods for planning itself, and they always involve some type of compromise.

Since the early days of AI, computing power has greatly improved and continues to improve (as per Moore's Law; see the end of the chapter for more information). This means larger state spaces can be searched much faster than ever before. However, there is still a limit to what can be done REAL TIME in *real time*, the time in which a physical robot moves around in a dynamic environment.

In general, a real robot cannot afford to just sit and deliberate. Why?

Deliberative, planner-based architectures involve three steps that need to be performed in sequence:

1. Sensing (S)

2. Planning (P)

3. Acting (A), executing the plan.

SPA
(SENSE-PLAN-ACT)
ARCHITECTURES

For this reason, deliberative architectures, which are also called *SPA (sense-plan-act) architectures*, shown in figure 13.3, have serious drawbacks for robotics. Here is what they are.

Drawback 1: Time-Scale

As we have said above, it can take a very long time to search in large state spaces. Robots typically have collections of sensors: some simple digital sensors (e.g., switches, IRs), some more complex ones (e.g., sonars, lasers, cameras), some analog sensors (e.g., encoders, gauges). The combined inputs from these sensors in themselves constitute a large state space. When combined with internal models or representations (maps, images of locations, previous paths, etc.), the result is a state space that is large and slow to search.

If the planning process is slow compared with the robot's moving speed, it has to stop and wait for the plan to be finished, in order to be safe. So to make progress, it is best to plan as rarely as possible and move as much as possible in between. This encourages open loop control (see Chapter 10), which we know is a bad idea in dynamic environments. If planning is fast, then execution need not be open loop, since replanning can be done at each step; unfortunately, typically this is impossible for real-world problems and robots.

Generating a plan for a real environment can be very slow.

Drawback 2: Space

It may take a great deal of space (i.e., memory storage) to represent and manipulate the robot's state space representation. The representation must contain all information needed for planning and optimization, such as distances, angles, snapshots of landmarks, views, and so on. Computer memory is comparatively cheap, so space is not as much of a problem as time, but all memory is finite, and some algorithms can run out of it.

Generating a plan for a real environment can be very memory-intensive.

Drawback 3: Information

The planner assumes that the representation of the state space is accurate and up to date. This is a reasonable assumption, because if the representation is not accurate and updated, the resulting plan is useless. For example, if a junction point in the maze is blocked, but the robot's internal map of the maze does not show it as blocked, the planned path might go through that junction and would therefore be invalid, but the robot would not discover that until it physically got to the junction. Then it would have to backtrack to a different path. The representation used by the planner must be updated and checked as often as necessary to keep it sufficiently accurate for the task. Thus, the more information the better.

Generating a plan for a real environment requires updating the world model, which takes time.

Drawback 4: Use of Plans

In addition to all of the above, any accurate plan is useful only if:

- The environment does not change during the execution of the plan in a way that affects the plan

- The robot knows what state of the world and of the plan it is in at all times

- The robot's effectors are accurate enough to execute each step of the plan in order to make the next step possible.

Executing a plan, even when one is available, is not a trivial process.

All of the above challenges of deliberative, SPA control became obvious to the early roboticists of the 1960s and 1970s, and they grew increasingly dissatisfied. As we learned in Chapter 2, in the early 1980s they proposed alternatives: reactive, hybrid, and behavior-based control, all of which are in active use today.

What happened to purely deliberative systems?

As a result of robotics work since the 1980s, purely deliberative architectures are no longer used for the majority of physical robots, because the combination of real-world sensors, effectors, and time-scale challenges outlined above renders them impractical. However, there are exceptions, because some applications demand a great deal of advance planning and involve no time pressure, while at the same time presenting a static environment and low uncertainty in execution. Such domains of application are extremely rare, but they do exist. Robot surgery is one such application; a perfect plan is calculated for the robot to follow (for example, in drilling the patient's skull or hip bone), and the environment is kept perfectly static (literally by attaching the patient's body part being operated on to the operating table with bolts), so that the plan remains accurate and its execution precise.

Pure deliberation is alive and well in other uses outside of robotics as well, such as AI for game-playing (chess, Go, etc.). In general, whenever the world is static, there is enough time to plan, and there is no state uncertainty, it is a good thing to do.

What about robotics problems that require planning?

The SPA approach has not been abandoned in robotics; it has been expanded. Given the fundamental problems with purely deliberative approaches, the following improvements have been made:

- Search/planning is slow, so save/cache important and/or urgent decisions

- Open loop plan execution is bad, so use closed loop feedback, and be ready to respond or replan when the plan fails.

In Chapter 15 we will see how the SPA model is currently incorporated into modern robots in a useful way.

To Summarize

- Deliberative architectures are also called SPA architectures, for sense-plan-act.

- They decompose control into functional modules which perform different and independent functions (e.g., sense-world, generate-plan, translate-plan-into-actions).

- They execute the functional modules sequentially, using the outputs of one as inputs of the next.

- They use centralized representation and reasoning.

- They may require extensive, and therefore slow, reasoning computation.

- They encourage open loop execution of the generated plans.

Food for Thought

- Can you use deliberative control without having some internal representation?

- Can animals plan? Which ones, and what do they plan?

- If you had perfect memory, would you still need to plan?

Looking for More?

- The Robotics Primer Workbook exercises for this chapter are found here: http://roboticsprimer.sourceforge.net/workbook/Deliberative_Control

- In 1965, Gordon Moore, the co-founder of Intel, made the following important observation: the number of transistors per square inch on integrated circuits had doubled every year since the integrated circuit was invented. Moore predicted that this trend would continue for the foreseeable future. In subsequent years, the pace slowed down a bit, to about every eighteen months. So Moore's Law states that the number of transistors per square inch on integrated circuits will double about every eighteen months until at least the year 2020. This law has an important impact on the computer industry, robotics, and even global economics.

14 *Don't Think, React!*
Reactive Control

Reactive control is one of the most commonly used methods for robot control. It is based on a tight connection between the robot's sensors and effectors.

REACTIVE SYSTEMS Purely *reactive systems* do not use any internal representations of the environment, and do not look ahead at the possible outcomes of their actions: they operate on a short time-scale and react to the current sensory information.

Reactive systems use a direct mapping between sensors and effectors, and minimal, if any, state information. They consist of collections of rules that couple specific situations to specific actions, as shown in figure 14.1. You can think of reactive rules as being similar to *reflexes*, innate responses that do not involve any thinking, such as jerking back your hand after touching a hot stove. Reflexes are controlled by the neural fibers in the spinal cord, not the brain. This is so that they can be very fast; the time it takes for a neural signal to travel from the potentially burned finger that touched the hot stove to the brain and back, and the computation in between to decide what to do takes too long. To ensure a fast reaction, reflexes don't go all the way to the brain, but only to the spinal cord, which is much more centrally located to most areas of the body. Reactive systems are based on exactly the same principle: complex computation is removed entirely in favor of fast, stored precomputed responses.

Reactive systems consist of a set of situations (stimuli, also called conditions) and a set of actions (responses, also called actions or behaviors). The situations may be based on sensory inputs or on internal state. For example, a robot may turn to avoid an obstacle that is sensed, or because an internal clock indicated it was time to change direction and go to another area. These examples are very simple; reactive rules can be much more complex, involving arbitrary combinations of external inputs and internal state.

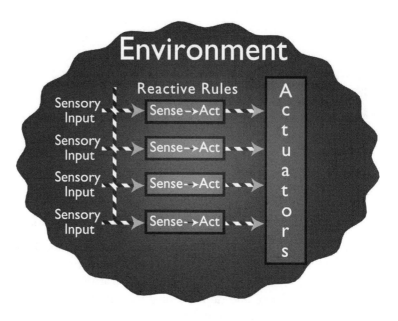

Figure 14.1 A diagram of a reactive architecture, showing the parallel, concurrent task-achieving modules.

The best way to keep a reactive system simple and straightforward is to have each unique situation (state) that can be detected by the robot's sensors trigger only one unique action of the robot. In such a design, the conditions are said to be *mutually exclusive*, meaning they exclude one another; only one can be true at a time.

MUTUALLY EXCLUSIVE CONDITIONS

However, it is often too difficult to split up all possible situations (world states) in this way, and doing so may require unnecessary encoding. To ensure mutually exclusive conditions, the controller must encode rules for all possible sensory input combinations. Recall from Chapter 3 that all those combinations, when put together, define the robot's sensor space. For a robot with a 1-bit such as a switch, that is total of two possibilities (on and off), for a robot with two switches, the total is four possibilities, and so on. As the sensors grow in complexity and number, the combinatorial space of all possible sensory inputs, i.e., the sensor space, quickly becomes unwieldy. In computer science, AI, and in robotics, the formal term for "unwieldy" is *intractable*. To encode and store such a large sensor space would take a giant

INTRACTABLE

lookup table, and searching for entires in such a giant table would be slow, unless some clever parallel lookup technique is used.

So to do a complete reactive system, the entire state space of the robot (all possible external and internal states) should be uniquely coupled or mapped to appropriate actions, resulting in the complete control space for the robot.

The design of the reactive system, then, is coming up with this complete set of rules. This is done at "design-time," not at "run-time," when the robot is active. This means it takes a lot of thinking by the designer (as any robot design process should), but no thinking by the robot (in contrast to deliberative control, where there is a lot of thinking done by the system).

In general, complete mappings between the entire state space and all possible responses are not used in manually-designed reactive systems. Instead, the designer/programmer identifies the important situations and writes the rules for those; the rest are covered with default responses. Let's see how that is done.

Suppose that you are asked to write a reactive controller that will enable a robot to move around and avoid obstacles. The robot has two simple whiskers, one on the left and one on the right. Each whisker returns 1 bit, "on" or "off"; "on" indicates contact with a surface (i.e., the whisker is bent). A simple reactive controller for wall-following using those sensors would look like this:

```
If left whisker bent, turn right.
If right whisker bent, turn left.
If both whiskers bent, back up and turn to the left.
Otherwise, keep going.
```

In the above example there are only four possible sensory inputs over all, so the robots' sensor space is four, and therefore there are four reactive rules. The last rule is a default, although it covers only one possible remaining case.

> *A robot using the above controller could oscillate if it gets itself into a corner where the two whiskers alternate in touching the walls. How might you get around this rather typical problem?*

There are two popular ways:

1. *Use a little randomness:* When turning, choose a random angle instead of a fixed one. This introduces variety into the controller and prevents it from getting permanently stuck in an oscillation. In general, adding a bit

Figure 14.2   Sonar ring configuration and the navigation zones for a mobile robot.

of randomness avoids any permanent stuck situation. However, it could still take a long time to get out of a corner.

2. *Keep a bit of history:* Remember the direction the robot turned in the previous step (1 bit of memory) and turn in the same direction again, if the situation occurs again soon. This keeps the robot turning in one direction and eventually gets it out of the corner, instead of getting it into an oscillation. However, there are environments in which this may not work.

Now suppose that instead of just two whiskers, your robot has a ring of sonars (twelve of them, to cover the 360-degree span, as you learned in Chapter 9). The sonars are labeled from 1 to 12. Sonars 11, 12, 1 and 2 are at the front of the robot, sonars 3 and 4 are on the right side of the robot, sonars 6 and 7 are in the back, and sonars 1 and 10 are on the left; sonars 5 and 8 also get used, don't worry. Figure 14.2) illustrates where the sonars are on the body of the robot. That is a whole lot more sensory input than two whiskers, which allows you to create a more intelligent robot by writing a whole lot more reactive rules.

But just how many rules do you need to have, and which ones are the right ones?

You could consider each sonar individually, but then you would have a lot of combinations of sonars to worry about and a lot of possible sonar values (from 0 to 32 feet each, recalling the range of Polaroid sonars from Chapter 9). In reality, your robot does not really care what each of the sonars individually returns; it is more important to focus on any really short sonar values (indicating that an obstacle is near) and specific areas around the robot (e.g., the front, the sides, etc.).

Let's start by defining just two distance areas around the robot:

1. Danger-zone: short readings, things are too close

2. Safe-zone: reasonable readings, good for following edges of objects, but not too far from things

Now let's take those two zones and write a reactive controller for the robot which considers groups of sonars instead of individual sonars:

```
(case
  (if (minimum (sonars 11 12 1 2))      ⇐ danger-zone
        and
        (not stopped)
  then
    stop)
  (if ((minimum (sonars 11 12 1 2))     ⇐ danger-zone
        and
        stopped)
  then
    move backward)
  (otherwise
    move forward))
```

The controller above consists of two reactive rules. As we have learned, it typically takes more than a single rule to get something done.

The controller stops the robot if it is too near an object in front of it, detected as the shortest reading of the front four sonars being in the danger-zone. If the robot is already stopped, and the object is too near (in the danger-zone), it backs up. The result is safe forward movement.

The controller does not check the sides and the back of the robot, however. If we assume that the environment the robot is in is static (furniture, walls) or that it contains rational people who keep their eyes open and do not run into the robot from the side or back, this controller will do just fine. But if the environment includes entities that are not very safe (children, other robots, hurried people), then it is in fact necessary to check all the sonars, not just the front four.

Moving forward and stopping for obstacles is not all we want from our robot, so let's add another controller, which we can consider another layer or module in a reactive architecture, to make it better able to get around its environment:

```
(case
  (if ((sonar 11 or 12) <= safe-zone
     and
     (sonar 1 or 2) <= safe-zone)
   then
     turn left)
  (if (sonar 3 or 4) <= safe-zone
   then
     turn right))
```

The above controller makes the robot turn away from detected obstacles. Since safe-zone is larger than danger-zone, this allows the robot to turn away gradually before getting too close to an obstacle and having to be forced to stop, as in the previous controller. If obstacles are detected on both sides, the robot consistently turns to the left, to avoid oscillations.

By combining the two controllers above we get a wandering behavior which avoids obstacles at a safe distance while moving smoothly around them, and also avoids collisions with unanticipated nearby obstacles by stopping and backing up.

14.1 Action Selection

The controllers described above use specific, mutually exclusive conditions (and defaults), so that their outputs are never in conflict, because only a single unique situation/condition can be detected at a time. If the rules are not triggered by mutually exclusive conditions, more than one rule can be triggered by the same situation, resulting in two or more different action commands being sent to the effector(s).

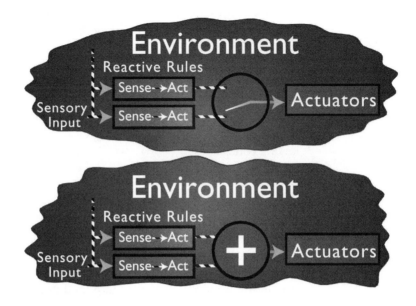

Figure 14.3   Two basic types of action selection: arbitration (top) and fusion (bottom).

ACTION SELECTION
Action selection is the process of deciding among multiple possible actions or behaviors. It may select only one output action or may combine the actions to produce a result. These two approaches are called arbitration and fusion.

COMMAND ARBITRATION
Command arbitration is the process of selecting one action or behavior from multiple candidates.

COMMAND FUSION
Command fusion is the process of combining multiple candidate actions or behaviors into a single output action/behavior for the robot.

The two alternatives are shown in figure 14.3 as part of the generic reactive system.

It turns out that action selection is a major problem in robotics, beyond reactive systems, and that there is a great deal of work (theory and practice) on different methods for command arbitration and fusion. We will learn a great deal more about this interesting problem in Chapter 17.

Although a reactive system may use arbitration to decide which action to execute, and thus execute only one action at a time, it still needs to monitor its rules in parallel, concurrently, in order to be ready to respond to any one or more of them that might be triggered.

Using the example of the wall-following robot, both whiskers have to be monitored all the time, or all sonars have to be monitored all the time, since an obstacle could appear anywhere, at any time. In the wall-following example, there are very few rules (at least in the whisker case), so given a fast processor, those could be checked sequentially and no real response time would be lost. But some rules may take time to execute. Consider the following controller:

```
If there is no obstacle in front, move forward.
If there is an obstacle in front, stop and turn away.
Start a counter.  After 30 seconds, choose randomly
    between left and right and turn by 30 degrees.
```

What kind of behavior does this controller produce? It produces random wandering with obstacle avoidance (and not very good avoidance at that). Note that the conditions of the three rules above require different amounts of time to compute. The first two check sensory input, while the third uses a timer. If the controller executes the rules sequentially, it would have to wait for 30 seconds in the third rule, before being able to check the first rule again. In that time, the robot could run into an obstacle.

Reactive systems must be able to support parallelism, the ability to monitor and execute multiple rules at once. Practically, this means that the underlying programming language must have the ability to *multitask*, to execute several processes/rules/commands in parallel. The ability to multitask is critical in reactive systems: if a system cannot monitor its sensors in parallel, and instead checks them in sequence, it may miss an event, or at least the onset of an event, and thus fail to react in time.

MULTITASKING

You can begin to see that designing a reactive system for a robot can be quite complicated, since multiple rules, potentially a large number of them, have to be put together in a way that produces effective, reliable, and goal-driven behavior.

> *How do we go about organizing a reactive controller in a principled way? By using architectures that have been specifically designed for that purpose.*

The best known architecture for reactive control is Subsumption Architecture, introduced by Prof. Rodney Brooks at MIT in 1985. It's an oldie by now, but it is still a goodie, finding its way into a vast number of reactive, hybrid, and behavior-based systems.

14.2 Subsumption Architecture

The basic idea behind Subsumption Architecture is to build systems incrementally, from the simple parts to the more complex, all the while using the already existing components as much as possible in the new stuff being added. Here is how it works.

Subsumption systems consist of a collection of modules or layers, each of which achieves a task. For example, they might be move-around, avoid-obstacles, find-doors, visit-rooms, pick-up-soda-cans, and so on. All of the task-achieving layers work at the same time, instead of in sequence. This means the rules for each of them are ready to be executed at any time, whenever the right situation presents itself. As you recall, this is the premise of reactive systems.

The modules or layers are designed and added to the robot incrementally. If we number the layers from 0 up, we first design, implement, and debug layer 0. Let's suppose that layer 0 is move-around, which keeps the robot going. Next we add layer 1, avoid-obstacles, which stops and turns or backs away whenever an obstacle is detected. This layer can already take advantage of the existing layer 0, which moves the robot around, so that, together, layers 0 and 1 result in the robot moving around without running into things it can detect. Next we add layer 2, say find-doors, which looks for doors while the robot roams around safely. And so on, until all of the desired tasks can be achieved by the robot through the combination of its layers.

But there is a twist. Higher layers can also temporarily disable one or more of those below them. For example, avoid-obstacles can stop the robot from moving around. What would result is a robot that sits still, but can turn and back away if somebody approaches it. In general, disabling obstacle avoidance is a dangerous thing to do, and so it is almost never done in any real robot system, but having higher layers selectively disable lower layers within a reactive system is one of the principles of Subsumption Architecture. This manipulation is done in one of only two ways, as shown in figure 14.4:

1. The inputs of a layer/module may be suppressed; this way the module receives no sensory inputs, and so computes no reactions and sends no outputs to effectors or other modules.

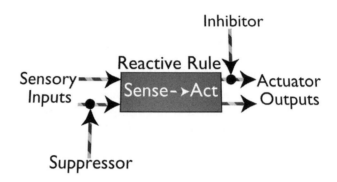

Figure 14.4   Subsumption methods for layer/module interaction: suppression of inputs (left) and inhibition of outputs (right).

2. The outputs of a layer/module may be inhibited; this way the module receives sensory inputs, performs its computation, but cannot control any effectors or other modules.

SUBSUMPTION ARCHITECTURE   The name *"Subsumption Architecture"* comes from the idea that higher layers can assume the existence of the lower ones and the goals they are achieving, so that the higher layers can use the lower ones to help them in achieving their own goals, either by using them while they are running or by inhibiting them selectively. In this way, higher layers "subsume" lower ones.

There are several benefits to the subsumption style of organizing reactive systems. First, by designing and debugging the system incrementally, we avoid getting bogged down in the complexity of the overall task of the robot. Second, if any higher-level layers or modules of a subsumption robot fail, the lower-level ones will still continue to function unaffected.

BOTTOM-UP   The design of subsumption controllers is called *bottom-up*, because it progresses from the simpler to the more complex, as layers are added incrementally. This is good engineering practice, but its original inspiration came to the inventor of Subsumption Architecture from biology. Brooks was inspired by the evolutionary process, which introduces new abilities based on the existing ones. Genes operate using the process of mixing (crossover) and

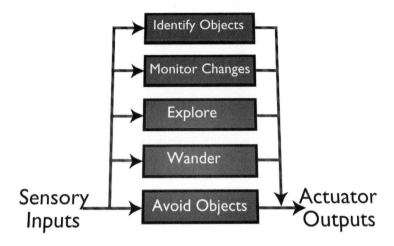

Figure 14.5 A subsumption-based robot control system.

changing (mutation) of the existing genetic code, so complete creatures are not thrown out and new ones created from scratch; instead, the good stuff that works is saved and used as a foundation for adding more good stuff, and so complexity grows over time. Figure 14.5 shows an example subsumption control system; it would be constructed bottom-up, adding each layer incrementally.

Building incrementally helps the design and debugging process, and using layers is useful for modularizing the robot controller. This is also good engineering practice. Otherwise, if everything is lumped together, it is hard to design, hard to debug, and hard to change and improve later.

The effectiveness of modularity also depends on not having all modules connected to all others, as that defeats the purpose of dividing them up. So in Subsumption Architecture, the goal is to have very few connections between different layers. The only intended connections are those used for inhibition and suppression. Inside the layers, of course, there are plenty of connections, as multiple rules are put together to produce a task-achieving behavior. Clearly, it takes more than one rule to get the robot to avoid obstacles, to find doors, and so on. But by keeping the rules for the separate tasks apart, the system becomes more manageable to design and maintain.

Therefore, in Subsumption Architecture, we use *strongly coupled* connections within layers, and *loosely coupled* connections between layers.

> *How do we decide what makes a subsumption layer, and what should go higher or lower?*

Unfortunately, there are no fixed answers to those hard design questions. It all depends on the specifics of the robot, the environment, and the task. There is no strict recipe, but some solutions are better than others, and most of the expertise that robot designers have, is gained through trial and error, basically through sweating it.

14.3   Herbert, or How to Sequence Behaviors Through the World

> *How would you make a reactive robot execute a sequence of behaviors? For example, consider the following task: Search for soda cans, find any empty ones, pick them up, and bring them back to the starting point.*

If you are thinking that you can have one layer or module activate another in a sequence, you are right, but that is not the way to go. Coupling between reactive rules or subsumption layers need not be through the system itself, through explicit communication, but instead through the environment.

A well-known subsumption robot called Herbert, designed by Jonathan Connell (a PhD student of Brooks') in the late 1980s, used this idea to accomplish the task of collecting soda cans. Here is how. Herbert had a layer that moved it around without running into obstacles. (As you will find out, every mobile robot has such a layer/ability/behavior, and it is usually reactive, for the obvious reason that collision-free navigation requires being able to respond quickly.) Herbert's navigation layer used infra red sensors. It also had a layer that used a laser striper and a camera to detect soda cans (by matching them to a particular width in a visual image, quite a clever trick). Herbert also had an arm and a layer of control that could extend the arm, sense if there was a can in its gripper, close the gripper, and pull the arm in. Cleverly, all these arm actions were separate, and were activated not in a sequential, internal way but instead by sensing the environment and the robot directly, like this:

```
If you see something that looks like a soda can
  approach it
If it still looks like a soda can up close
  then extend the arm
  else turn and go away

Whenever the arm is extended
check between the gripper fingers

Whenever the gripper sensors (IR break beam) detects
something
  close the gripper

Whenever the gripper is closed and the arm extended
pull back the arm

Whenever the arm is pulled back and the gripper closed
go to drop off the can
```

Herbert had a very clever controller; see figure 14.6 for its diagram. It never used any "explicit" sequencing, but instead just checked the different situations in the world (the environment and the robot together) and reacted appropriately. Note, for example, that it was safe for Herbert to close its gripper whenever there was something sensed in it, because the only way it could sense something in the gripper was after it had already looked for cans and extended the arm to reach one. Similarly, it was safe for Herbert to pull the arm back when the gripper was closed, since it was closed only if it was holding a can. And so on.

By the way, in case you are wondering how it knew if the cans were empty or full, it used a simple strain gauge to measure the weight of the can it was holding. It took only the light cans.

Herbert had no internal communication between the layers and rules that achieved can-finding, grabbing, arm-retracting, and moving around, yet it executed everything in the right sequence every time, and was able to perform its can-collecting task. The big idea here is that Herbert was using the sensory inputs as a means of having its rules and layers interact. Brooks called this "interaction through the world."

Another motto, also by Brooks, recommends that we "use the world as its own best model." This is a key principle of Subsumption Architecture,

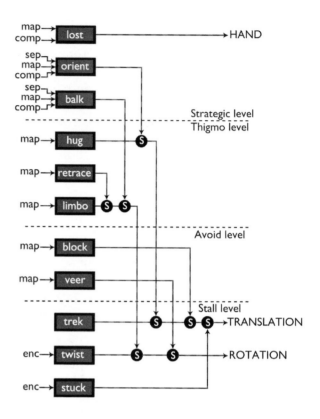

Figure 14.6   Herbert's control architecture.

and of reactive systems in general: If the world can provide the information directly (through sensing), it is best for the robot to get it that way rather than to store it internally in a representation, which may be large, slow, expensive, and outdated.

To summarize, the guiding principles of Subsumption Architecture are:

• Systems are built from the bottom up.

• Components are task-achieving actions/behaviors (not functional modules).

• Components can be executed in parallel (multitasking).

- Components are organized in layers.

- Lowest layers handle the most basic tasks.

- Newly added components and layers exploit the existing ones. Each component provides and does not disrupt a tight coupling between sensing and action.

- There is no use of internal models; "the world is its own best model."

Subsumption Architecture is not the only method for structuring reactive systems, but it is a very popular one, due to its simplicity and robustness. It has been widely used in various robots that successfully interact with uncertain, dynamically changing environments.

How many rules does it take to put together a reactive system?

By now, you know the answer is "That depends on the task, the environment, and the sensors on the robot." But it is important to remember that for any robot, task, and environment that can be specified *in advance*, a complete reactive system can be defined that will achieve the robot's goals. However, that system may be prohibitively large, as it may have a huge number of rules. For example, it is in theory possible to write a reactive system for playing chess by encoding it as a giant lookup table that gives the optimal move for each possible board position. Of course the total number of all possible board positions, and all possible games from all those positions, is too huge for a human to remember or even a machine to compute and store. But for smaller problems, such as tic-tac-toe and backgammon, this approach works very well. For chess, another approach is needed: using deliberative control, as we learned in Chapter 13.

To Summarize

- Reactive control uses tight couplings between perception (sensing) and action to produce timely robotic response in dynamic and unstructured worlds (think of it as "stimulus-response").

- Subsumption Architecture is the best-known reactive architecture, but certainly not the only one.

- Reactive control uses a task-oriented decomposition of the controller. The control system consists of parallel (concurrently executed) modules that achieve specific tasks (avoid-obstacle, follow-wall, etc.).

- Reactive control is part of other types of control, specifically hybrid control and behavior-based control.

- Reactive control is a powerful method; many animals are largely reactive.

- Reactive control has limitations:

  - Minimal (if any) state

  - No memory

  - No learning

  - No internal models / representations of the world.

Food for Thought

- Can you change the goal of a reactive system? If so, how? If not, why not? You will soon learn how other methods for control deal with this problem and what the trade-offs are.

- Can you always avoid using any representation/world model? If so, how? If not, why not, and what could you do instead?

- Can a reactive robot learn a maze?

Looking for More?

- The Robotics Primer Workbook exercises for this chapter are found here: http://roboticsprimer.sourceforge.net/workbook/Reactive_Control

- Genghis, the six-legged robot we mentioned in Chapter 5 was also programmed with Subsumption Architecture. You can learn about how its control system was put together, how many rules it took, and how it gradually got better at walking over rough terrain as more subsumption layers were added, from *Cambrian Intelligence* by Rodney Brooks.

15 Think and Act Separately, in Parallel
Hybrid Control

As we have seen, reactive control is fast but inflexible, while deliberative control is smart but slow. The basic idea behind hybrid control is to get the best of both worlds: the speed of reactive control and the brains of deliberative control. Obvious, but not easy to do.

HYBRID CONTROL *Hybrid control* involves the combination of reactive and deliberative control within a single robot control system. This combination means that fundamentally different controllers, time-scales (short for reactive, long for deliberative), and representations (none for reactive, explicit and elaborate world models for deliberative) must be made to work together effectively. And that, as we will see, is a tall order.

In order to achieve the best of both worlds, a hybrid system typically consists of three components, which we can call layers or modules (though they are not the same as, and should not be confused with, layers/modules used in reactive systems):

- A reactive layer

- A planner

- A layer that links the above two together.

As a result, hybrid architectures are often called *three-layer architectures* and hybrid systems, *three-layer systems*. Figure 15.1 is a diagram of a hybrid architecture and its layers.

We already know about deliberation with planners and about action with reactive systems. What we now need to learn about is the real challenge and "value added" of hybrid systems: the "magic middle." The middle layer has a hard job, because it has to:

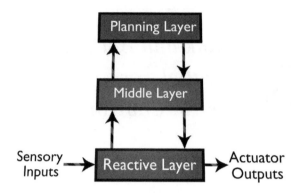

Figure 15.1   A diagram of a hybrid architecture, showing the three layers.

- Compensate for the limitations of both the planner and the reactive system

- Reconcile their different time-scales

- Deal with their different representations

- Reconcile any contradictory commands they may send to the robot.

So the main challenge of hybrid control is achieving the right compromise between the deliberative and reactive parts of the system.

Let's work through an example to see what that means in practice. Suppose we have a robot whose job is to deliver mail and paperwork through the day to various offices within an office building. This was a very popular mobile robot application that researchers liked to think about in the 1970s, 1980s, and 1990s, until email became the preferred means of communication. However, office buildings still have a great many tasks that require "gophering" or running around with deliveries. Also, we can use the same controller to deliver medicines to patients in a hospital, so let's consider that example.

In order to get around a busy office/hospital environment, the robot needs to be able to respond to unexpected obstacles, and fast-moving people and objects (gurneys, stretchers, rolling lunch carts, etc.). For this it needs a robust reactive controller.

In order to efficiently find specific offices/rooms for making deliveries, the robot needs to use a map and plan short paths to its destinations. For this it needs an internal model and a planner of some kind.

So there you have it, a perfect candidate for a hybrid system. We already know how to do both of the component systems (reactive collision-free navigation and deliberative path planning), so how hard can it be to put the two together?

Here is how hard:

• What happens if the robot needs to deliver a medication to a patient as soon as possible, yet it does not have a plan for a short path to the patient's room? Should it wait for the plan to be computed, or should it move down the corridor (in which direction?) while it is still planning in its head?

• What happens if the robot is headed down the shortest path but suddenly a crew of doctors with a patient on a stretcher starts heading its way? Should it just stop and get out of the way in any direction, and wait as long as it may take, or should it start replanning an alternative path?

• What happens if the plan the robot computed from the map is blocked, because the map is out of date?

• What happens if the patient was moved to another room without the robot knowing?

• What happens if the robot keeps having to go to the same room, and so has to replan that path or parts of it all the time?

• What if, what if, what if.

The methods for handling the above situations and various others are usually implemented in the "magic middle" layer of a hybrid system. Designing the middle layer and its interactions with the other two layers (as shown in figure 15.2) are the main challenges of hybrid control. Here are some commonly used methods and approaches.

15.1 Dealing with Changes in the World/Map/Task

When the reactive system discovers that it cannot do its job (for example, it can't proceed because of an obstacle, a closed door, or some other snag), it can

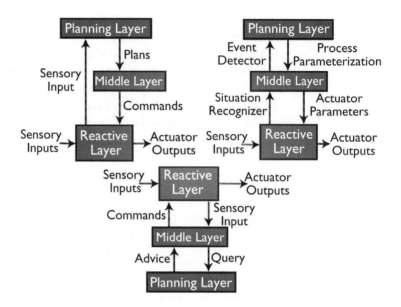

Figure 15.2 Some ways of managing layer interaction in hybrid controllers for robots.

inform the deliberative layer about this new development. The deliberative layer can use this information to update its representation of the world, so that it can now, and in the future, generate more accurate and useful plans.

This is a good idea not only because it is necessary to update the internal model when things change, but also because we already know that updating internal models and generating plans takes time and computation, so it cannot be afforded continually. The input from the reactive layer provides an indication of a very good time for such an update.

15.2   Planning and Replanning

Whenever the reactive layer discovers that it cannot proceed, this can be used as a signal to the deliberative layer to do some thinking, in order to generate a new plan. This is called *dynamic replanning*.

But not all information flows from the bottom up, from the reactive to the deliberative layer. In fact, the deliberative layer, which provides the path to the goal (i.e., the room to go to), gives the robot the directions to follow, turns to take, distances to go. If the planner is computing while the robot is moving, it may send a message to the reactive navigation layer to stop, turn around, and head in a different direction because a better way has been discovered in the map.

In general, a complete plan, when finished, has the best answer the deliberator can generate. But sometimes there is not enough time to wait for that full and optimal answer (such as when the patient really needs immediate help). In those cases, it is often best to get the robot going in the generally right direction (for example, toward the correct wing of the hospital) and in the meantime keep generating a more accurate and detailed plan, and updating the navigation layer as needed.

Of course the timing of these two processes, the reactive navigation (running around) and deliberative planning (thinking hard), is obviously not synched up, so the robot may get to the right wing of the hospital and not know where to go next, and thus may have to stop and wait for the planner to get done with its next step. Alternatively, the planner may need to wait for the robot to move around or get out of a crowded area in order to figure out exactly where it is, so that it can generate a useful plan. We will learn more about the various challenges of navigation in Chapter 19.

15.3   Avoiding Replanning

A useful idea researchers in planning had quite a while ago was to remember/save/store plans so that they would not have to be generated again in the future. Of course, every plan is specific to the particular initial state and goal state, but if those are at all likely to recur, it is worth stashing away the plan for future use.

This idea is really popular for situations that happen often and need a fast decision. In our example above, such situations include dealing with crowded junctions of corridors or dynamic obstacles that interfere with a particular direction of movement. Instead of "thinking out" how to deal with this every time, controller designers often preprogram many of these responses ahead of time. But because these are not quite reactive rules, because they involve multiple steps, and yet they are not quite plans, because they do not involve very many steps, they find a perfect home in the middle layer of the three-layer system.

This basic idea of storing and reusing mini plans for repeated situations has been used in "contingency tables," literally lookup tables that tell the robot what to do (as in a reactive system) and pull out a little plan as a response (as in a deliberative system). This same idea has appeared in the form of fancy terms such as "intermediate-level actions" and "macro operators." In all cases, it is basically the same thing: plans that are computed (either off-line or during the robot's life) and stored for fast lookup in the future.

So who has ultimate priority, the deliberative layer or the reactive layer?

The answer is, as usual, "It depends." What it depends on includes the type of environment, task, sensing, timing, and reaction requirements, and so on. Some hybrid systems use a hierarchical structure, so that one of the systems is always in charge. In some cases, the deliberator's plan is "the law" for the system, while in others the reactive system merely considers it as advice that can be ignored. In the most effective systems, the interaction between thinking and acting is coupled, so that each can inform and interrupt the other. But in order to know who should be in charge when, it is necessary to consider the different modes of the system and specify who gets its say. For example, the planner can interrupt the reactive layer if it (the planner) has a better path than the current being executed, but it should do so only if the new plan is sufficiently better to make it worth the bother. On the other hand, the reactive layer can interrupt the planner if it (the reactive layer) finds a blocked path and cannot proceed, but should only do so after it has tried to get around the barrier for a while.

15.4   On-Line and Off-Line Planning

In everything we've talked about so far, we have assumed that the deliberation and the reaction are happening as the robot is moving around. But as we saw from the role of the middle layer, it is useful to store plans once they are generated. So the next good idea is to preplan for all the situations that might come up, and store those plans ahead of time. This *off-line planning* takes place while the robot is being developed and does not have much to worry about, as compared with *on-line planning* of the kind that a busy robot has to worry about while it is trying to get its job done and its goals achieved.

OFF-LINE PLANNING

ON-LINE PLANNING

If we can preplan, why not generate all possible plans ahead of time, store them all, and just look them up, without ever having to do search and deliberation at run-time, while the robot is trying to be fast and efficient?

This is the idea behind universal plans.

UNIVERSAL PLAN

A *universal plan* is a set of all possible plans for all initial states and all goals within the state space of a particular system.

If for each situation a robot has a preexisting optimal plan it only has to look up, then it can always react optimally, and so have both reactive and deliberative capabilities without deliberating at all. Such a robot is reactive, since the planning is all done off-line and not at run-time.

Another good feature of such precompiled plans is that information can be put into the system in a clean, principled way. Such information about

DOMAIN KNOWLEDGE

the robot, the task, and the environment is called *domain knowledge*. It is compiled into a reactive controller, so the information does not have to be reasoned about (or planned with) on line, in real time but instead becomes a set of real-time reactive rules that can be looked up.

This idea was so popular that researchers even developed a way of generating such precompiled plans automatically, by using a special programming language and a compiler. Robot programs written in that language produced a kind of a universal plan. To do so, the program took as input a mathematical specification of the world and of the robot's goals, and produced a control "circuit" (a diagram of what is connected to what) for a reactive "machine." These machines were called *situated automata*; they were not real, physical machines but formal (idealized) ones whose inputs were connected to abstract (again idealized, not real) sensors, and whose outputs were connected

SITUATED AUTOMATA

to abstract effectors. To be *situated* means to exist in a complex world and to interact with it; *automata* are computing machines with particular mathematical properties.

Unfortunately, this is too good to be true for real-world robots. Here is why:

- The state space is too large for most realistic problems, so either generating or storing a universal plan is simply not possible.

- The world must not change; if it does, new plans need to be generated for the changed environment.

- The goals must not change; this is the same as with reactive systems, where if the goals change, at least some of the rules need to change as well.

Situated automata could not be situated in the real, physical world after all. Robots exist in the real world but we can't easily write mathematical specifications that fully describe this world, which is what that the situated automata approach required.

And so we are back to having to do deliberation and reaction in real time, and hybrid systems are a good way to do that. They do have their own drawbacks, of course, including:

- The middle layer is hard to design and implement, and it tends to be very special-purpose, crafted for the specific robot and task, so it has to be reinvented for almost every new robot and task.

- The best of both worlds can end up being the worst of both worlds; if mismanaged, a hybrid system can degenerate into having the planner slow down the reactive system, and the reactive system ignore the planner entirely, minimizing the effectiveness of both.

- An effective hybrid system is not easy to design or debug, but that is true for any robot system.

In spite of the drawbacks of hybrid systems, they are the most popular choice for a great many problems in robotics, especially those involving a single robot that has to perform one or more tasks that involve thinking of some type (such as delivery, mapping, and many others) as well as reacting to a dynamic environment.

To Summarize

- Hybrid control aims to bring together the best aspects of reactive control and deliberative control, allowing the robot to both plan and react.

- Hybrid control involves real-time reactive control in one part of the system (usually the low level) and more time-expensive deliberative control in another part of the system (usually the high level), with an intermediate level in between.

- Hybrid architectures are also called three-layer architectures because of the three distinct layers or components of the control system.

- The main challenge of hybrid systems lies in bringing together the reactive and deliberative components in a way that results in consistent, timely, and robust behavior over all.

- Hybrid systems, unlike reactive systems, are capable of storing representation, planning, and learning.

Food for Thought

Is there an alternative to hybrid systems, or is that all there is in terms of intelligent, real-time robot behavior? Can you think of another way that a robot can be able to both think and react? After you come up with an answer, check out the next chapter.

Looking for More?

- The Robotics Primer Workbook exercises for this chapter are found here: http://roboticsprimer.sourceforge.net/workbook/Hybrid_Control

- Hybrid control is a major area of research in control theory, a field of study usually pursued in electrical engineering, which we described in Chapter 2. (Recall that control theory covers feedback control, which we studied in Chapter 10.) In control theory, the problem of hybrid control deals with the general interplay between continuous and discrete signals and representations, and the control of systems that contain them. Here are a couple of popular textbooks on control theory:

    - *Signals and Systems* by Alan V. Oppenheim, Alan S. Willsky, and with Nawab S. Hamid.
    - *Modern Control Systems*, by Richard C. Dorf and Robert H. Bishop.

16 *Think the Way You Act Behavior-Based Control*

As we have learned, reactive control and deliberative control each have their limitations, and hybrid control, an attempt to combine their best components, has its own challenges. In this chapter we will learn about *behavior-based control*, another popular way of controlling robots, which also incorporates the best of reactive systems but does not involve a hybrid solution.

Behavior-based control (BBC) grew out of reactive control, and was similarly inspired by biological systems. Actually, if you think about it, all approaches to control (reactive, deliberative, hybrid, and behavior-based) were inspired by biology, at one level or another. This just goes to show that biological systems are so complex that they can serve as inspiration for a variety of disparate methods for control, yet still remain more complicated and effective than anything artificial that has been made so far. Just when you start to feel really good about artificial systems, all you need to do is to go outside and look at some bugs to see how far we have yet to go.

Back to behavior-based systems. The primary inspiration came from several main challenges:

- Reactive systems are too inflexible, incapable of representation, adaptation, or learning.

- Deliberative systems are too slow and cumbersome.

- Hybrid systems require complex means of interaction among the components.

- Biology seems to have evolved complexity from simple and consistent components.

If you think about the different control methodologies along a line, BBC is closer to reactive control than to hybrid control, and farthest from delibera-

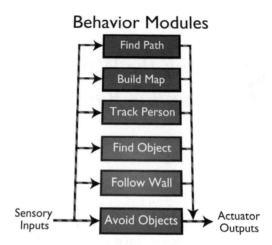

Figure 16.1   A diagram of a behavior-based architecture, showing behavior organization.

tive control. In fact, as you will see, behavior-based systems have reactive components, just as hybrid systems do, but they do not have traditional deliberative components at all. Figure 16.1 shows a diagram of a generic behavior-based control architecture.

So what is behavior-based control?

BEHAVIOR-BASED CONTROL

Behavior-based control (BBC) involves the use of "behaviors" as modules for control. Thus, BBC controllers are implemented as collections of behaviors. The first property of behavior-based control to remember is that it's all about behaviors. That brings up the obvious question:

What is a behavior?

There is a definite answer to that question, and in fact one of the strengths of BBC comes from different ways in which people have encoded and implemented behaviors, which are also sometimes called *behavior-achieving modules*. But don't assume that anything goes, and any piece of code can be a behavior. Fortunately, there are some rules of thumb about behaviors and constraints on how to design them and what to avoid in implementing them:

- Behaviors achieve and/or maintain particular goals. A *homing* behavior achieves the goal of getting the robot to the home location. A *wall-following* behavior maintains the goal of following a wall.

- Behaviors are time-extended, not instantaneous. That means they take some time to achieve and/or maintain their goals. After all, it takes a while to go home or follow a wall.

- Behaviors can take inputs from sensors and also from other behaviors, and can send outputs to effectors and to other behaviors. This means we can create networks of behaviors that "talk to" each other.

- Behaviors are more complex than actions. While a reactive system may use simple actions like *stop* and *turn-right*, a BBC uses time-extended behaviors like the ones we saw above, as well as others like *find-object, follow-target, get-recharged, hide-from-the-light, aggregate-with-your-team, find-mate,* etc.

LEVEL OF ABSTRACTION

As you can see from the short list above, behaviors can be designed at a variety of levels of detail or description. This is called their *level of abstraction*, because *to abstract* is to take details away and make things less specific. Behaviors can take different amounts of time and computation. In short, they are quite flexible, and that is one of the key advantages of BBC.

The power and flexibility of BBC comes not only from behaviors but also from the organization of those behaviors, from the way they are put together into a control system. Here are some principles for good BBC design:

- Behaviors are typically executed in parallel/concurrently, much as in reactive systems, in order to enable the controller to respond immediately when needed.

- Networks of behaviors are used to store state and to construct world models/representations. When assembled into distributed representations, behaviors can be used to store history and to look ahead into the future.

- Behaviors are designed so that they operate on compatible time-scales. This means it is not good BBC design to have some very fast behaviors and some very slow ones. Why not? Because that makes the system hybrid in terms of the time-scale, and we've already seen (in Chapter 15) that interfacing different time-scales is a challenging problem.

Because of the above properties of behaviors and their combinations, BBC is not limited in the ways that reactive systems are. At the same time, it does not employ a hybrid structure, as hybrid systems do. Instead, behavior-based systems have the following key properties:

1. The ability to react in real-time

2. The ability to use representations to generate efficient (not only reactive) behavior

3. The ability to use a uniform structure and representation throughout the system (with no intermediate layer(s)).

Before we move on to the other advantages of BBC, let's clear something up first. You might be wondering about the difference between behaviors inside the robot and inside a behavior-based controller, which cannot be observed from the outside, and behaviors that the robot performs and that can be observed from the outside. If you are not wondering about this, you should be. Here is why.

In some behavior-based systems, the internal behavior structure exactly matches the externally manifested behaviors. This means that if the robot performs wall-following, then there is a distinct piece of the control program we can call "wall-following" that achieves that observable behavior. However, this is not always the case, especially for more complex behaviors, so most controllers are not designed this way.

Why not? After all, having an intuitive match between the internal control program and the externally observable behavior sounds really clean and easy to understand. Unfortunately, it turns out to be impractical.

Most interesting observable robot behaviors are a result not just of the control program, but also of the interaction of the internal behaviors among themselves and with the environment the robot is in. This is basically the same idea that reactive systems used (recall Chapter 14): simple reactive rules can interact to produce interesting observable behavior of the robot. The same is true for internal control behaviors.

Consider flocking, the behavior in which a group of robots moves together in a group. A robot that flocks with others does not necessarily need to have an internal *flocking* behavior. In fact, flocking can be implemented very elegantly in a completely distributed way. To learn more about this, see Chapter 18.

This is one of the clever philosophies of behavior-based systems. Such systems are typically designed so the effects of the behaviors interact in the environment rather than internally through the system, in order to take advantage of *interaction dynamics*. In this context, those dynamics refer to patterns and history of interaction and change. The idea that rules or behaviors can interact to produce more complex outputs is called *emergent behavior*; we will devote entire Chapter 18 to it because it is quite confusing, yet interesting and important.

In general, in designing a behavior-based system, the designer begins by listing the desirable observable, externally manifested behaviors. Next, the designer figures out how best to program those behaviors with internal behaviors. Those internal behaviors, as we saw, may or may not directly achieve the observable goals. To make this process easier, various compilers and programming languages have been developed.

Now we know that behavior-based systems are structured as collections of internal behaviors, which, when moving in an environment, produce a collection of externally manifested behaviors, and the two are not necessarily the same. And that's all right, as long as the robot's observable behaviors achieve its overall goals.

Let's consider the following example problem: We need the robot to move around a building and water plants whenever they get dry.

You can imagine that the robot needs the following behaviors: avoid-collisions, find-plant, check-if-dry, water, refill-water-reservoir, recharge-batteries. To program such a robot, would you program those exact six behaviors and be done? Possibly, but chances are that some of them, especially the more complicated ones (like find-plant, for example), may themselves consist of internal behaviors themselves, such as wander-around, detect-green, approach-green, and so on. But don't assume that this means behavior-based systems are simply hierarchical, with each behavior having its own component behaviors. Some systems are hierarchical, but usually multiple behaviors use and share the same underlying component behaviors. In our example, the refill-water-reservoir behavior may also use the wander-around behavior in order to get back to the water supply and refill its reservoir.

In general, behavior-based controllers are networks of internal behaviors which interact (send messages to each other) in order to produce the desired external, observable, manifested robot behavior.

As we outlined earlier, the key differences among the various control methods are based on the way each treats modularity, time, and representation.

In the case of BBC, the approach to modularity is that of using a collection of behaviors, and having those behaviors be relatively similar in terms of execution time. This means that having one behavior which contains a centralized world model and performs reasoning on it, as in deliberative or hybrid systems, would not fit the behavior-based philosophy, and would thus not make a good controller.

Since BBC is based on the reactive philosophy (but not limited by it), it also mandates that behaviors be incrementally added to the system, and that they be executed concurrently, in parallel, rather than sequentially, one at a time. Behaviors are activated in response to external and/or internal conditions, sensory inputs, and internal state or messages from other behaviors. Dynamics of interaction arise both within the system itself (from the interaction among the behaviors) and within the environment (from the interaction of the behaviors with the external world). This is similar to the principles and effects of reactive systems, but can be exploited in richer and more interesting ways because:

> *Behaviors are more expressive (more can be done with them) than simple reactive rules are.*

Remember that behavior-based systems can have reactive components. In fact, many of these systems do not use complex representations at all. Internal representation is not necessary for all problems, and so it can be avoided for some. This does not mean that the resulting controller is reactive; as long as the controller is structured using behaviors, in the ways described above, it creates a BBC system, with all the benefits that come with the approach, even without the use of representation.

Because behaviors are more complex and more flexible than reactive rules, they can be used in clever ways to program robots. By having behaviors interact with each other within the robot, they can be used to store representation. Therefore they can serve as a basis for learning and prediction, and this means that behavior-based systems can achieve the same things that hybrid systems can, but in a different way. One of the major differences is in how representation is used.

16.1   Distributed Representation

The philosophy of behavior-based systems mandates that information used as internal representation not be centralized or centrally manipulated. This is in direct contrast to deliberative and hybrid systems, which typically use a centralized representation (such as a global map) and a centralized reasoning mechanism (such as a planner that uses the global map to find a path). What is an alternative to that approach, and why is it a good thing? That's what we are going to learn about next.

The key challenge in using representation in BBC is in how that representation (any form of world model) can be effectively distributed over the behavior structure. In order to avoid the pitfalls of deliberative control, the representation must be able to act on a time-scale that is close to (if not the same as) the real-time components of the system. Similarly, to avoid the challenges of the middle layer in hybrid control, the representation needs to use the same underlying behavior structure as the rest of the system.

16.2 An Example: Distributed Mapping

Let's return to our plant-watering robot. Suppose that your task is to improve that robot so it does not just move around randomly, look for plants, and water them, but also to make a map of its environment as it goes along, and then to use that map to find shortest paths to particular plants, or other locations in the environment.

Navigation and mapping are the most common mobile robot tasks, as we will learn in Chapter 19. So let's see how we can build (or learn) a map in a BBC. We already know that it shouldn't be a traditional centralized map that a deliberative or hybrid system would use.

> *Instead, we have to somehow distribute information about the environment over behaviors in a useful way. How can we do that?*

Here is how: we can distribute parts of a map over different behaviors. Then we can connect parts of the map that are adjacent in the environment (such as a wall that is next to a corridor) by connecting the behaviors that represent them in the robot's head/controller. This way we will build a network of behaviors which represents a network of locations in the environment. What is connected in the map is connected in the environment, and vice versa. A fancier formal way to say this is that the topology of the map is isomorphic (has the same form) with the topology of the specific landmarks in the environment.

Figure 16.2   Toto the robot. (Photo courtesy of the author)

16.2.1   Toto the Robot

There once was a robot named Toto (see Toto's photo in figure 16.2) that used just such a representation to learn the map of its office environment at MIT, in order to hang out with students and help someone[1] get her PhD in robotics. Let's see how Toto learned its environment and got around to do its task. To be honest, Toto did not in fact water any plants, as water and mobile robots rarely mix well.

16.2.2   Toto's Navigation

Toto was a behavior-based system; you can see its control diagram in figure 16.3. It was equipped with a simple ring of twelve Polaroid sonars (if you can't remember why twelve, go back to Chapter 14), and a low-resolution compass (2 bits). Toto lived around 1990, in the "old days," well before laser scanners became commonplace. As in any good BBC, Toto's controller consisted of a collection of behaviors. Its lowest levels of the system took care of moving Toto around safely, without collisions. Next, Toto had a behavior that kept it near walls and other objects, based on the distance perceived

---

1. Ok, I'll tell you who: it was this book's author.

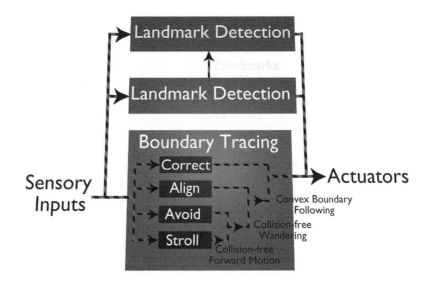

Figure 16.3 Toto's behavior-based controller.

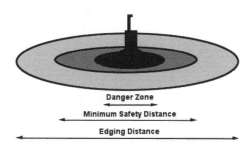

Figure 16.4 Toto's sonar-based navigation regions. (Image courtesy of the author)

with its sonar sensors. Just as we learned in Chapter 14, Toto used a set of sonar sensing regions for navigation (shown in figure 16.4); its low-level safe navigation was in fact a reactive system. In fact, Toto's navigation controller used some of the programs you saw in Chapter 14.

Following boundaries was a good idea for Toto. Remember why? If not, look back at Chapter 9 and recall that sonars are most likely to be accurate at short and intermediate distances from objects. It pays to get accurate sensor data when building maps and going places, so Toto did it by staying near objects whenever possible. It turns out that rats and mice also like to stick near objects, perhaps for other reasons (safety from people?), but nobody would have confused Toto with a rat, at least based on its looks. We'll see later that Toto's brain, in particular its use of distributed representation, was also rat-like, but more on that in a bit.

To summarize, Toto's navigation layer manifested the following behaviors:

1. The lowest-level behavior kept it moving safely around its environment without colliding with static obstacles or moving people.

2. The next level behavior kept it near walls and other objects in the environment.

16.2.3   Toto's Landmark Detection

The next thing Toto did was to pay attention to what it was sensing and how it was moving. This is generally a good thing to do. It allowed Toto to notice when it was moving straight in the same direction for a while, or when it was meandering around a lot. Moving straight indicated that it might be next to a wall or in a corridor, and so Toto was able to detect walls and corridors. Meandering around indicated that it was in a cluttered area, and so Toto would detect that as well. So while Toto moved about, one of its behaviors was paying attention to what its sonars "saw," and in that way detecting and recognizing landmarks: If Toto kept moving along in a consistent compass direction (in a near-straight line) and kept close to a wall on one side (say the left), it would recognize and detect a left-wall (LW). If the wall was on the other side, it detected a right-walls (RW), and if it saw walls on both sides, it recognized a corridor (C). Finally, if it detected nothing straight for a while (no walls on either side), causing it to meander about, it declared a messy-area (MA). In Toto's mind, left walls, right walls, corridors, and messy areas were special, because it could notice them easily from its sensors and its own behavior. Therefore, they formed nice *landmarks* for his map.

Toto's landmark detection layer also noted the compass direction for each of the landmarks it detected. Its 2-bit compass (that's not meant as an insult, it is an accurate description) indicated north as 0, east as 4, south as 8, west as 12, and so on, as shown in figure 16.5. In addition to the compass direction of the landmark, Toto also noted the landmark's approximate length.

To summarize, Toto's landmark detection layer manifested the following behavior:

1. It kept track of the consistency in the sonar and compass readings, and, noticing those, recognized one of the landmark types (left-wall, right-wall, corridor, messy-area).

2. For each landmark type, it also noted the compass direction and length of the landmark.

16.2.4 Toto's Mapping Behaviors

In order to store those landmarks, Toto constructed a map, as shown in figure 16.5. Here we come upon the really interesting fact about Toto: its distributed map representation. Each of the landmarks that Toto discovered was stored in its own behavior. When Toto's landmark detection layer discovered a corridor (C) going north (0) for 6.5 feet, it created a behavior that looked something like this:

```
my-behavior-type:  C
my-compass-direction:  0
my-approximate-location:  (x,y)
my-approximate-length:  6.5
whenever received (input)
  if input(behavior-type) = my-behavior-type
    and
  input(compass-direction) = my-compass-direction
  then
    active <- true
```

What you see above is pseudo-code, not really Toto's code but something close to it and easier to read. The code above simply says that the behavior represents a corridor (C) going north (0), so that whenever the input into that behavior matches that description (i.e., when the input is C and 0, and the

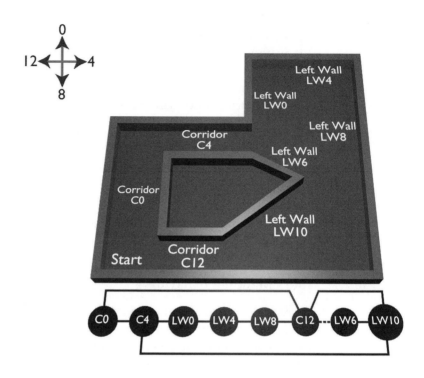

Figure 16.5   An example of an indoor environment Toto navigated in, and a map it constructed of that environment.

approximate length and location are right), the landmark matches and the robot is in fact in that particular corridor.

Toto's landmarks were clever: they worked regardless of which way Toto was going down a corridor or along a wall. A C0 shown above would also match C8 (because that corridor goes north-south, so Toto might be heading down the same corridor in the opposite direction), and C4 would also match C12 (because that corridor goes east-west). The same rule applied to walls as well: a left wall going north (LW0) matched a right wall going South (RW8), and so on.

Whenever the landmark-detecting layer detected a landmark, its description (type and compass direction) was sent to *all* map behaviors at the same time (in parallel). If any behavior in the map matched the input, as described above, then that behavior would become *active*, meaning Toto knew where it was in its map; this is called *localization* and is a very important part of the navigation problem. We will talk about it in detail in Chapter 19.

Ok, so far we have Toto moving around safely, detecting landmarks, and reporting those to the behavior map. Whenever one matches, Toto recognizes that it is in a place where it has been before.

What happens if no behavior in the map matches the landmark Toto is detecting?

That means Toto has discovered a new place/landmark, which needs to be added to the map. Suppose Toto is seeing a corridor (C) going east (4) that is approximately 5 feet long. Here is how it adds that new landmarks/behavior to its map: it takes a brand new, empty behavior "shell" and assigns it to the newly found landmark information, so it becomes:

```
my-behavior-type:  C
my-compass-direction:  4 or 12
my-approximate-location:  (x,y)
my-approximate-length:  5.0
whenever received (input)
  if input(behavior-type) = my-behavior-type
    and
  input(compass-direction) = my-compass-direction
  then
    active <- true
```

It then connects the previous behavior (suppose it is the C0 we just saw earlier) to the new one by putting a communication link between them. This means that when Toto is going north on C0, the next landmark it will see is C4.

To take advantage of knowing where it is (being localized), Toto's map did another clever thing: the behavior that was active sent messages to its neighbor in the direction in which Toto was traveling (for example, in the map shown in figure 16.5, C0 sent messages to C4 if Toto was going north, but to C12 if Toto was going south), to warn it that it should be next to be active. If the neighbor was next to be recognized, then Toto was even more confident about the accuracy of its map and its localization. On the other hand, if the expecting neighbor was not recognized next, that usually meant that Toto had branched off on a new path it had not tried before, and would soon discover new landmarks to be added to the map.

Besides sending messages to its neighbors to tell them who is coming next, Toto's map behaviors sent inhibitory messages to one another. Specifically, when a behavior matches the incoming landmark, it sends messages to all of its neighbors to inhibit them, so that only one map behavior can be active at one time, since Toto can really be only in one place at one time. That's another part of Toto's localization. (If this seems like a lot of machinery to assure Toto where it is, you'll see in Chapter 19 how hard it is for any robot to know where it is.)

To summarize, Toto's mapping layer performed the following behaviors:

1. Matched the received landmark against all landmark behaviors in the map

2. Activated the map behavior that matched the current landmark

3. Inhibited the previous landmark

4. Reported if no landmark matched, and created a new landmark behavior in response, storing the new landmark and connecting it to the previous landmark.

16.2.5 Path Planning in Toto's Behavior Map

As we have seen, while Toto roamed around its environment, it built up a landmark map of its world. This is an illustration of how you can build a distributed map with a behavior-based system. Now let's see how such a map is used to find/plan paths.

Let's suppose that Toto's watering can was empty and it needed to go to the corridor with the water supply, so this landmark became Toto's goal. If the goal landmark and Toto's current position were the same behavior in the map, then Toto was already at the goal and did not need to go anywhere. That's too lucky to have been the case most of the time. Instead, Toto had to plan a path between its current location and the goal. Here is how it did that.

The behavior that corresponded to the goal sent messages saying "Come this way!" to all of its neighbors. The neighbors, in turn, sent the messages on to their own neighbors (but not back the way the message came; that would be a waste and create some problems). Each map behavior, when passing a message along, also added to it its landmark length, so that as the path went

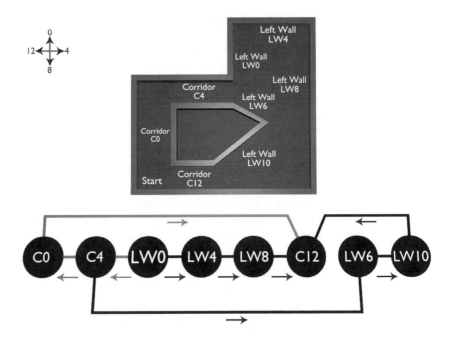

Figure 16.6   Toto's way of path planning.

through the network, its length grew based on the number and lengths of map behaviors it passed through.

Rather quickly, these messages all reached the currently active behavior which represented wherever Toto was at the time. When they arrived, the messages contained the total length of the path taken to get there, in terms of physical landmark length. The current landmark paid attention only to the message with the minimum summed landmark length, since that indicated the direction toward the shortest path to the goal.

Besides going to a particular landmark, such as a specific corridor, Toto could also find the nearest landmark with a particular property. For example, suppose that Toto needed to find the nearest right wall. To make this happen, all right wall landmarks in the map would start sending messages. Toto followed the shortest path, and reached the nearest right wall in the map.

16.2.6   Toto's Map-Following

As we learned in Chapter 13, the environment can change at any point, the robot can get lost or confused about where it is, and so on, and therefore planning a path once and sticking to in an open loop fashion is a dangerous prospect. So Toto did not plan once, but kept planning continually. Here is how.

Toto's map behaviors sent messages *continuously*, so wherever in the map Toto happened to be (whatever behavior happened to be active) it could respond to the shortest path it received. This is a very useful property if the environment is likely to change while the robot is moving around. Another reason for such continual updating of the plan is the possibility of the so-called *kidnapped robot problem*, in which somebody picks up the robot and moves it to another location. Continuously recomputing the path by sending the messages around means that a blocked path or changed location will immediately be replaced by a better (shorter) path in the network. In fact, it is very important to notice that Toto did not really have any notion of a "path" at all! Instead, it just went from one landmark to the next, which eventually led it to the goal, as efficiently as possible within its map. It did not compute the path explicitly or store it anywhere, but instead used the *active behavior map* to find its way, no matter where it was.

KIDNAPPED ROBOT PROBLEM

Toto's way of building a map and planning within its active structure is not easy to describe, but it is easy (efficient) to compute. It is also not easy to illustrate, but figure 16.6 tries to do so, by showing how messages are passed between neighboring behaviors in Toto's map in order to plan a path between its current behavior and goal behavior.

To summarize, Toto's path finding layer performed the following behaviors:

1. Repeatedly sent (one or more) messages from the goal throughout the map network, until the goal was reached

2. Kept a running total of the length of the path as the messages were passed through the network

3. At each landmark Toto reached (and thus recognized and activated), the current total of incoming messages indicated where Toto should go next (to what behavior in the map), telling it how to turn and where to head.

As a result of all of its behaviors operating in parallel and interacting with each other and with the environment, Toto was able to:

1. Move around safely and quickly

2. Detect new and known landmarks

3. Build a map and find its position in the map

4. Find shortest paths toward various goals.

All of the above was achieved in real time, so no detectable time-scale issues arose. Additionally, no intermediate layer existed in Toto because there was no need to explicitly interface any layers at different time-scales (e.g., navigation vs. path finding). Toto is a good example of a complete behavior-based system capable of achieving both real-time behavior and optimization without the use of a hybrid controller.

This example, albeit somewhat complicated, shows us that BBC and hybrid control are equally powerful and expressive. By using distributed representations, we can do arbitrary tasks with BBC, just as with hybrid systems.

How does one decide when to use a hybrid controller and when to use a behavior-based one?

There is no fixed answer to that question; the decision is often based on the designer's personal preferences and past experience and expertise. In some domains, there are advantages to one approach or the other, however. For example, for single-robot control, especially for complex tasks involving representation and planning, hybrid systems are typically most popular. In contrast, for controlling groups and teams of robots, behavior-based control is often preferred, as we will learn in Chapter 20.

Some fans of behavior-based control claim that BBC is the most realistic model of biological cognition (thinking), because it includes the spectrum of computational capabilities, from simple reflexes to distributed representations, all within a consistent time-frame and representation. Other fans prefer BBC for pragmatic reasons, finding it easy to program and use for developing robust robots. (Of course, those who prefer hybrid systems often list the very same reasons for their opposite choice!)

To Summarize

- Behavior-based systems (BBS) use behaviors as the underlying modularity and representation.

- BBS enable fast real-time responses as well as the use of representation and learning.

- BBS use distributed representation and computation over concurrent behaviors.

- BBS are an alternative to hybrid systems, with equal expressive power.

- BBS require some expertise to program cleverly, as any control approach does.

Food for Thought

- Some people say that behaviors should be more precisely defined so that behavior-based robot programming would be easier to figure out. Others believe that having behaviors be loosely defined allows robot programmers to be creative and come up with interesting and novel ideas for ways to design robot controllers. What do you think?

- The field of psychology went through a phase called "behaviorism", which defined all organisms based on their observable behaviors. This general idea is similar to behavior-based robotics. However, behaviorism believed that there was no difference between externally observable behaviors (actions) and internally observable behaviors (thinking and feeling). Some behaviorists did not even believe in internal state (recall that from Chapter 3). As we saw in this chapter, BBS do have internal state and representation, so they are not consistent with behaviorism. People often confuse behaviorism and behavior-based systems. Is the difference clear to you? If not, you might want to read more about both.

Looking for More?

- The Robotics Primer Workbook exercises for this chapter are found here: http://roboticsprimer.sourceforge.net/workbook/Behavior-Based_Control

- You can learn more about Toto from the following papers:

  Matarić, Maja J. (1990), "A Distributed Model for Mobile Robot Environment-Learning and Navigation", MIT EECS Master's Thesis, Jan 1990, *MIT AI Lab Tech Report AITR-1228*, May.

  Matarić, Maja J. (1992), "Integration of Representation Into Goal-Driven Behavior-Based Robots", *IEEE Transactions on Robotics and Automation*, 8(3), June, 304-312.

- *Cambrian Intelligence* by Rodney A. Brooks is a great collection of papers about behavior-based robots built at MIT during the 1980s and 1990s, including Genghis, Attila, Toto, and several others.

- After finishing this book, Ronald Arkin's *Behavior-Based Robotics* is an excellent next step in learning about robotics.

17 *Making Your Robot Behave*
Behavior Coordination

BEHAVIOR
COORDINATION

Any robot that has at its disposal more than one behavior or action has to solve the surprisingly difficult problem of *action selection* or *behavior coordination*.

> *The problem is simple on the surface: what action/behavior should be executed next?*

Solving this problem in a way that makes the robot do the right thing over time, so as to achieve its goals, is anything but simple. Both hybrid and behavior-based systems have multiple modules, and therefore have to solve this coordination problem. In hybrid systems, the middle layer is often stuck with the job, while in behavior-based systems the job may be distributed throughout the entire control system.

In general, there are two basic ways of selecting the next behavior or the next action: picking one behavior/action or combining multiple behaviors/actions. As we mentioned in Chapter 14, these two approaches are called arbitration and fusion. Figure 17.1 shows some of the many methods available for implementing each and combinations of the two.

17.1 Behavior Arbitration: Make a Choice

ARBITRATION

Arbitration is the process of selecting one action or behavior from multiple possible candidates. We have already discussed arbitration in the context of reactive control, in Chapter 14, but it has a much broader role to play in robotics in general, since it applies any time there is a choice to be made among different actions or behaviors.

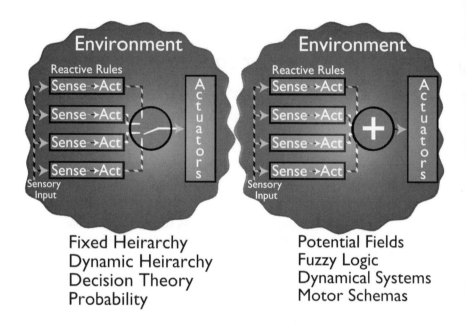

Fixed Heirarchy
Dynamic Heirarchy
Decision Theory
Probability

Potential Fields
Fuzzy Logic
Dynamical Systems
Motor Schemas

Figure 17.1   Various methods for behavior coordination.

Arbitration-based behavior coordination is also called *competitive* behavior coordination, because multiple candidate behaviors compete but only one can win.

Arbitration can be done with:

- A fixed priority hierarchy (behaviors have preassigned priorities)

- A dynamic hierarchy (behavior priorities change at run-time).

Subsumption Architecture (from Chapter 14) used a fixed priority hierarchy of behaviors, which was implemented through inhibition of the behaviors' outputs and suppression of their inputs.

Many hybrid systems employ fixed priority hierarchies of control. As we saw in Chapter 15, in some systems the planner always has control over the reactive system, and in others the reverse is always the case.

More sophisticated hybrid systems use dynamic hierarchies in order to have the control of the system transition between the reactive and deliberative parts of the controller, for best performance. Similarly, behavior-based systems often employ dynamic arbitration to decide what behavior wins and takes control of the robot next.

Dynamic arbitration usually involves computing some function to decide who wins. The function can be anything, including voting (so the winner is "elected"), fuzzy logic, probability, or spreading of activation, among many others.

17.2 Behavior Fusion: Sum It Up

In contrast to behavior arbitration, a competitive method of control, there are various architectures that employ the execution of multiple behaviors at the same time. This is the basis of behavior fusion.

BEHAVIOR FUSION *Behavior fusion* is the process of combining multiple possible candidate actions or behaviors into a single output action/behavior for the robot. Behavior fusion is also called a "cooperative" method because it combines outputs of multiple behaviors to produce the final result, which may be one or more of the existing behaviors, or even a wholly new result. (How can that be? Read on, you'll see.)

Combining behaviors, however, is a very tricky business. Even walking and chewing gum at the same time is sometimes a challenge, and things only get more difficult from there. (Robots are no good with gum, and they are not great at walking yet, either.)

In order to combine two or more behaviors, the behaviors must be represented in a way that makes that combination possible and, hopefully, easy. Here is an example of how that could be done.

Consider a robot that has the following behaviors:

```
Stop:
command velocity 0 to both wheels

GoForward:
command velocity V1 to both wheels
```

```
TurnRightABit:
command velocity V2 to the right wheel,
and 2*V2 to the left wheel

TurnLeftABit:
command velocity V3 to the left wheel,
and 2*V3 to the right wheel

TurnRightALot:
command velocity V4 to the right wheel,
and 3*V4 to the left wheel

TurnLeftALot:
command velocity V5 to the left wheel,
and 3*V5 to the right wheel

SpinInPlaceToTheRight:
command velocity V6 to the left wheel,
and -V6 to the right wheel

SpinInPlaceToTheLeft:
command velocity V7 to the right wheel,
and -V7 to the left wheel
```

The control fusion mechanism takes the outputs of all of the behaviors, sums the velocity commands for each wheel, and sends the result to each wheel.

Do you see any potential problems with this approach?

It is certainly nice and simple, and since it uses only velocities, all values can be directly added and subtracted. But what happens if two behaviors cancel each other out? What if V2 and V3 are equal, and TurnRightABit and TurnLeftABit are commanded at the same time? What will the robot do? It will not turn at all! Worse yet, if we combine the Stop behavior with any other, the robot will not stop, and that could be a very bad thing if there is an obstacle in front of it.

The most obvious basic case in which this problem arises is around obstacles. Consider the following (popular) basic controller:

```
GoToGoal:
compute the direction toward the goal
    go in that direction
AvoidObstacle:
compute the direction away from the obstacle
    go in that direction
```

Together, these two behaviors can make the robot oscillate in front of obstacles, doing a little dance and never making progress.

How might we get around some of these problems?

It turns out that roboticists have thought about this long and hard, and there is a great deal of theory and even more practice showing what works and what doesn't. For example, some behaviors could be weighted so that their outputs have a stronger influence on the final result. Of course, that will not always work either. Another possibility is to have some logic in the system which prevents certain combinations and certain outcomes. This makes the otherwise nice and simple controller not so nice and simple any more, but it is often necessary. There is no free lunch.

It is in general easy to do command fusion when the commands are some numerical value, such as velocity, turning angle, and so on. Here is an example from "the street":

Consider the DAMN (Dynamic Autonomous Mobile Navigation) Architecture, which was used to control a van that autonomously drove on roads over very large distances. (DAMN was developed by Julio Rosenblatt at CMU; see the end of the chapter for details.) The demo was called "No Hands Across America" and got a lot of well-deserved publicity. In the DAMN controller, various low-level actions, such as turning angles, were all voting members. Each action got to vote (for itself or others), all votes were tallied, and the result, a weighted sum of actions, was executed. Figure 17.2 shows a diagram of the control architecture. This is a clear example of command fusion and it turned out to be a very effective way of controlling the driving robot. Later on, it was improved further through the use of fuzzy logic, another way to fuse commands. You can find more information about fuzzy logic at the end of the chapter.

Formal methods for command fusion include potential fields, motor schemas, and dynamical systems, to name but a few. These are all well-developed

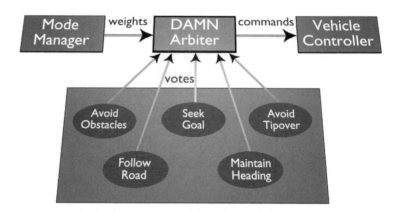

Figure 17.2 A diagram of the DAMN architecture for dynamic autonomous mobile navigation (thus the name).

and deserve careful study. We'll skip them here but you can find pointers to additional reading at the end of this chapter.

In general, fusion methods are applied to lower-level representations of actions or behaviors (direction, velocity), while arbitration is used for higher-level ones (turn, stop, grasp). Many control systems use a combination of methods, with fusion at one level and arbitration at another.

To Summarize

- Deciding what action, behavior, or set of behaviors to execute is a complex problem in robotics.

- This problem is called *behavior coordination*.

- Behavior coordination has been studied for a long time and there are a great many possible solutions. However, no solution is perfect and there are no strict rules about what is the right solution for a given robot system, environment, and task.

Food for Thought

Does it ever happen to you that you can't decide between two options? What do you do? Theories from cognitive science and neurosciences say Does it ever happen to you that you can't decide between two options? What do you do? Theories from cognitive science and neurosciences say that human emotions are useful for helping us decide between similar options; there exist people with certain neural deficits who are both unemotional and unable to choose between things that appear similarly good (or bad). Does this mean robots need emotions? Some researchers in AI have argued that both robots and AI software in general could benefit from such emotions. What do you think?

Looking for More?

- Ronald Arkin's *Behavior-Based Control* gives an easy-to-understand explanations for some of the more popular methods for behavior fusion, including schemas and potential fields.

- Julio Rosenblatt's PhD dissertation was titled *DAMN: A Distributed Architecture for Mobile Navigation*; you can get it as a technical report from CMU. In general, you can get any PhD dissertation from the institution that issued it.

- Paolo Pirjanian's PhD dissertation *Multiple Objective Action Selection & Behavior Fusion using Voting* from Aalborg University in Denmark provides an excellent overview of behavior coordination approaches.

- Fuzzy logic is a rather popular topic with a great many books published on the subject. You might want to start with *Fuzzy Logic: Intelligence, Control and Information* by John Yen and Reza Langari.

18 *When the Unexpected Happens Emergent Behavior*

The whole idea that a robot can produce unexpected behavior is somewhat scary, at least to a non-roboticist.

But is any robot behavior really unexpected? And is all unexpected behavior bad?

In this chapter we will learn about emergent behavior, an important but poorly understood phenomenon.

All robot behavior results from the interaction of the robot's controller and the environment. If either or both of these components are simple, the resulting behavior may be quite predictable. On the other hand, if the environment is dynamic, or if the controller has various interacting components, or both, the resulting robot behavior may be surprising, in a good or bad sense. If such unexpected behavior has some structure, pattern, or meaning to an observer, it is often called emergent. But hang on, we won't define emergent behavior yet; we are not quite ready for that.

18.1 An Example: Emergent Wall-Following

Although it is easy to imagine how unexpected behavior results from complicated systems, you should not assume that it takes a complex environment or a complex controller to generate emergent behavior. For a counterexample, consider the following simple controller (which should by now be quite familiar to you):

```
If left whisker bent, turn right.
If right whisker bent, turn left.
If both whiskers bent, back up and turn to the left.
```

Figure 18.1 Emerging wall-following behavior by a simple mobile robot.

```
Otherwise, keep going.
```

What happens if you put a robot with this controller next to a wall? See figure 18.1 for the answer. It follows the wall! And yet, the robot itself knows nothing about walls, since the controller has nothing explicit about walls in its rules. No walls are detected or recognized, yet the robot follows walls reliably and repeatably. In fact, the controller above might have been written simply for obstacle avoidance, but it follows walls, too.

So is wall-following an emergent behavior of this controller?

Well, that depends.

What is emergent behavior, and is it magic?

18.2 The Whole Is Greater Than the Sum of Its Parts

EMERGENT BEHAVIOR The intuitive if rather imprecise notion of *emergent behavior* is that it involves some type of "holistic" capability, where the behavior of the robot is greater than the sum of its "parts," which are the rules in the controller. Emergent behavior appears to produce more than what the robot was programmed to do. Somehow we get more than we built in, something for nothing, something magic.

As you saw in Chapters 14 and 16, interactions between rules, behaviors, and the environment provide a source of richness and expressive power for system design, and good roboticists often exploit this potential to design clever and elegant controllers. Specifically, reactive and behavior-based systems are often designed to take advantage of just such interactions; they are designed to manifest what some might call emergent behaviors.

Let's consider another example. Suppose you have a group of robots and you need to make them flock together. Suppose further that you cannot have the robots communicate with each other and must have each robot just follow its own rules and its own local sensory data (just as birds do when they flock). How would you program each robot so as to have the group flock?

It turns out that such distributed flocking can be achieved with the following set of simple rules for each of the robots:

```
Don't get too close to other robots (or other obstacles)
Don't get too far from other robots
Keep moving if you can
```

What we set "close" and "far" depends on various parameters, including the velocities of the robots, their sensor range, and on how tight we want the flock to be.

When run in parallel, the rules above will result in the group of robots flocking. And furthermore, if you want to get the flock to go to a particular place, such as South (useful for birds) or Home (useful for everyone), you need to add only one more rule, which "nudges" each robot to the desired destination, assuming the rest of the rules above are satisfied. To learn more about this type of flocking, wait for Chapter 20 and also look into the resources at the end of this chapter.

What does it take to create an emergent behavior?

18.3   Components of Emergence

The notion of emergence depends on two components:

1. The existence of an external observer, to see the emergent behavior and describe it (because if nobody sees the emergent behavior, how do we know it has happened?)

2. Access to the innards of the controller, to verify that the behavior is not explicitly specified anywhere in the system (because if it is, then it is not emergent, merely expected).

Many roboticists use the combination of the two components above as the definition of emergent behavior. The wall-following and flocking behaviors described above work quite nicely for this definition.

Of course, not all researchers agree with that definition (such ready agreement in a research field would be just too easy and too boring). Some argue that there is nothing emergent about the wall following and flocking behaviors, and that they are merely examples of a particular style of indirect robot programming that uses the environment and side-effects.

These critics propose that in order for a behavior to be emergent, it must be truly unexpected; it must be discovered "surreptitiously" by observing the system, and must come as a true surprise to the external observer.

18.4   Expect the Unexpected

The problem with the unexpected surprise as a property of a behavior is that it depends entirely on the expectations of the observer, and those expectations are completely subjective. A naive observer is surprised by many observations. A prehistoric man would be very surprised indeed by a bicycle staying upright or a plane flying. (In fact, the only reason most of us are not surprised by these phenomena is from familiarity; it is certainly not because most of us actually understand exactly why bicycles and airplanes do not fall.) In contrast, an informed observer may not be surprised by much, if he or she understands the system extremely well or has seen it in action before.

This last point above brings us to an even trickier problem with unexpected emergent behavior. Once such a behavior is observed, it is no longer unexpected. Just as with the bicycle and airplane, even a caveman would eventually get used to seeing them upright or in the sky, and would expect them to stay so, even without understanding how the system works. And that lack of surprise – the fulfilled expectation, the predictability – invalidates the second definition of emergence. But this is no good, because a behavior either is or isn't emergent; it should not be emergent once (the first time it is seen), and then never again.

18.5 Predictability of Surprise

So, it seems pretty clear that the knowledge of the observer cannot be a valid measure of emergence, and that something more objective is needed. A more objective and concrete way to try to define and understand emergence is to look a bit more closely at the property of the controller "not knowing" about the behavior and yet generating it.

As we saw in the wall-following example, the controller (and thus the robot) had no notion of "walls" or "following." Wall-following emerged from the interaction of the simple avoidance rules and the structure of the environment. If there were no walls in the environment, wall-following would not emerge. This brings us to another definition of emergent behavior:

EMERGENT BEHAVIOR
Emergent behavior is structured (patterned, meaningful) behavior that is apparent at one level of the system (from the observer's viewpoint), but not at another (from the controller's/robot's viewpoint).

This means that if the robot generates interesting and useful behavior without being explicitly programmed to so, the behavior can be called emergent, whether or not it comes as a surprise to any observer.

Some such behaviors, such as wall-following and flocking examples, can be preprogrammed through clever rules and are not particularly surprising to the designer, but are still clever ways to do robot control. However, a great deal of emergent behavior cannot be designed in advance, and is indeed unexpected, produced by the system as it runs and emerging from the dynamics of interaction that arise only at execution time. In such cases, the system has to be run and observed in order for the behavior to manifest itself; it cannot be predicted.

> *Why is that? Why can't we, in theory, predict all possible behaviors of the system (assuming we are very smart)?*

Well, we could in theory predict all possible behaviors of a system, but it would take forever, because we would have to consider all possible sequences and combinations of the robot's actions and behaviors in all possible environments. And to make it even more complicated, we would also have to take into account the unavoidable uncertainty of the real world, which affects all of the robot's sensing and actions.

It is easy to misunderstand and misinterpret these ideas. The fact that we cannot predict *everything* in advance does not mean we cannot predict *anything*. We can make certain guarantees and predictions for all robots and environments and tasks. We can, for example, predict, and thus guarantee the performance of a surgical robot. If we couldn't, who would use it? Still, we cannot be sure that there may not be an earthquake while it operates, which may well interfere with its performance. On the other hand, for some controllers and robots we may not, for example, be able to predict what robot will put what brick where but can guarantee that the robots will, together, build a wall (unless of course somebody steals all the bricks or there is an earthquake, again).

And so, we are left with the realization that the potential for emergent behavior exists in any complex system, and since we know that robots that exist in the real world are, by definition, complex systems, emergence becomes a fact of life in robotics. Different roboticists approach it in different ways: some work hard to avoid it, others aim to use it, and still others try to ignore it whenever possible.

18.6 Good vs. Bad Emergent Behavior

Notice that so far we have assumed that all emergent behavior is, in some way, desirable and good. Of course all sorts of structured behaviors can emerge that are not at all desirable, but instead quite problematic for the robot. Oscillations are one obvious case in point; a robot with simple turn-away-from-the-obstacle rules can easily get stuck in a corner for a while. Can you think of other examples?

> *Can you think of how to make an emergent behavior come about without programming it in directly?*

Furthermore, for any kind of emergent behavior:

> *What is the difference between an emergent bug and an emergent feature?*

That is entirely dependent on the observer and his or her goals. One observer's bug is another's feature. If you want your robot to get stuck in corners, then oscillations are a feature; otherwise they are a bug.

Now let's get back to purposefully creating emergent behavior in our robots.

Recalling our examples of wall-following and flocking, we can see that emergent behavior can arise from the interactions of the robot and the environment over time and/or over space. In wall following, the rules (go left, go right, keep going), in the presence of a wall, result over time in following the wall. So what is necessary, besides the controller itself, is the structure in the environment and enough time for the behavior to emerge.

In contrast, in the flocking behavior, what is necessary is the many robots executing the behaviors together, in parallel. This is again a type of environment structure, albeit in the form of other robots. The rules themselves (don't go too close, don't go too far, keep going) mean nothing if the robot is alone, and in fact it takes three or more robots for such flocking to be stable; otherwise it falls apart. But given enough robots, and enough space and time for them to move, a flock emerges.

> *Given the necessary structure in the environment, and enough space and/or time, there are numerous ways in which emergent behaviors can be programmed and exploited.*

This is often particularly useful in programming teams of robots; see Chapter 20 to learn more.

18.7 Architectures and Emergence

As you would expect, the different control architectures affect the likelihood of generating and using emergent behavior in different ways, because system modularity directly affects emergence.

Reactive and behavior-based systems employ parallel rules and behaviors, respectively, which interact with each other and the environment, thus providing the perfect foundation for exploiting emergent behavior by design. In contrast, deliberative systems are sequential, and thus typically have no parallel interactions between the components, and thus would require environment structure in order to have any behavior emerge over time. Similarly, hybrid systems follow the deliberative model in attempting to produce a coherent, uniform output of the system, minimizing interactions and thus minimizing emergence.

To Summarize

- Emergent behavior can be expected or unexpected, desirable or undesirable.

- Roboticists try to avoid the undesirable but exploit the desirable emergent behavior.

- Emergent behavior requires an observer of the system's behavior and information about the insides of the system

- Different control architectures have different methods for exploiting or avoiding emergent behavior.

- You can spend a lot of years studying emergent behavior and philosophizing about it.

- People, the ultimate complex systems situated in complex environments, produce emergent behavior, too.

Food for Thought

- Besides wall-following, or following (servoing to some target), which is a natural behavior to produce via emergence, can you think of others you might want to implement this way?

- Can you think of ways of using emergent behavior to do something quite complicated, such as build a bridge or a wall or a building? Ants do it all the time, see Chapter 20.

Looking for More?

- The Robotics Primer Workbook exercises for this chapter are found here: http://roboticsprimer.sourceforge.net/workbook/Emergent_Behavior

- Craig Reynolds invented "boids", simulated bird-like creatures that were programmed with simple rules and produced elegant flocking behaviors. You can read about boids in *Artificial Life* by Steven Levy. You can also find more information about boids, and watch videos of the flocks in action, on the Internet.

19 *Going Places*
Navigation

NAVIGATION *Navigation* refers to the way a robot finds its way in the environment.

Getting from one place to another is remarkably challenging for a robot. In general, you will find that any robot controller spends a great deal of its code getting where it needs to be at any given time, compared with its other "high-level" goals. Getting any body part where it needs to be is hard, and the more complicated the robot's body, the harder the problem.

The term "navigation" applies to the problem of moving the robot's whole body to various destinations. Although the problem of navigation has been studied in the domain of mobile robots (as well as flying and swimming robots), it applies to any robot that can move around. The body of the robot may be of any form; the locomotion mechanism takes care of moving the body appropriately, as we saw in Chapter 5, and the navigation mechanism tells it where to go.

What's so hard about knowing where to go?

As usual, the problem is rooted in uncertainty. Since a robot typically does not know exactly where it is, this makes it rather hard to know how to get to its next destination, especially since that destination may not be within its immediate sensory range. To better understand the problem, let's break it down into a few possible scenarios, in all of which the robot has to find some object, let's say a puck.

• Suppose the robot has a map of its world which shows where the puck is. Suppose also that the robot knows where it is in its world, the map. What remains to be done to get to the puck is to plan a path between the robot's current location and the goal (the puck), then follow that plan. This is the *path planning problem*. Of course if anything goes wrong - if the map is incorrect or the world changes - the robot may have to update the map, search around, replan, and so on. We already talked a bit about how to deal with such challenges of deliberative and hybrid control in Chapters 13 and 15.

• Now suppose the robot has a map of its world which shows where the puck is, but the robot does not know where it is in the map. The first thing the robot must do is find itself in the map. This is the *localization* problem. Once the robot localizes within its map, i.e., knows where it is, it can plan the path, just as above.

• Now suppose the robot has a map of its world and knows where it is in its map, but does not know where the puck is in the map (or world, same thing). What good is the map if the location of the puck is not marked? Actually, it is a good thing indeed. Here is why. Since the robot does not know where the puck is, it will have to go around searching. Given that it has a map, it can use the map to plan out a good searching strategy, one that covers every bit of its map and is sure to find the puck, if one exists. This is the *coverage problem*.

COVERAGE PROBLEM

• Now suppose the robot does not have a map of its world. In that case, it may want to build a map as it goes along; this is the *mapping* problem. Notice that not having a map does not mean the robot does not know where it is; you may not have a map of New York City, but if you are standing next to the Statue of Liberty, you know where you are: next to the Statue of Liberty. Basically, a robot's local sensors can tell it where it is if the location can be uniquely recognized, such as with a major landmark like the Statue of Liberty, or with a global positioning system (GPS). However, without a map you won't know how to get to the Empire State Building. To make the problem really fun, and also quite realistic for many real-world robotics domains, suppose the robot does not know where it is. Now the robot has two things it has to do: figure out where it is (localization) and find the puck (search and coverage).

• Now suppose that the robot, which has no map of its world and does not know where it is, chooses to build a map of its world as it goes along trying to localize and search for the puck. This is the *simultaneous localization and mapping (SLAM)* problem, also called concurrent mapping and localization (CML), but that term is not as catchy or popular. This is a "chicken or egg" problem: to make a map, you have to know where you are, but to know where you are, you have to have a map. With SLAM, you have to do both at the same time.

All of the various problems mentioned above are components of the navigation problem. To summarize, they are:

- Localization: finding out where you are

- Search: looking for the goal location

- Path planning: planning a path to the goal location

- Coverage: covering all of an area

- SLAM: localization and mapping at the same time.

As it turns out, every one of the above problems has been studied extensively in robotics, and numerous research papers, and even some books, have been written about them. Various algorithms exist for every one of the above problems, and more are developed every year by many eager robotics researchers. In short, all of the above are hard problems, still considered "open," meaning research is ongoing in trying to solve them in increasingly better ways. Given that, it is hard to do justice to the large body of work that exists in these areas of navigation. Next, we will briefly mention some of the approaches to the problems above, but to learn more, check out the additional reading sources given at the end of the chapter.

19.1 Localization

ODOMETRY

PATH INTEGRATION

One way to stay localized is through the use of odometry. *Odometry* comes from the Greek *hodos* meaning "journey" and *metros* meaning "measure," so it literally means measuring the journey. A more formal term for it is *path integration*. Cars have odometers, as do robots. Robot odometry is usually based on some sensors of the movement of the robot's wheels, typically shaft encoders, as we learned in Chapter 8. However, the farther the robot goes and the more turns it makes, the more inaccurate odometry becomes. This is unavoidable, since any measurement of a physical system is noisy and inaccurate, and any small error in the odometry will grow and accumulate over time. The robot will need to recognize a landmark or find some other way to reset its odometer if it is to have any chance of localizing. Odometry in robots is the same basic process that people use in dead reckoning.

Odometry allows the robot to keep track of its location relative to the starting or reference point. Therefore, it is localized relative to that point. In general, localization is always relative to some reference frame, whether it is an arbitrary starting point or a GPS map.

STATE ESTIMATION

To deal with localization in a more formal way, it is usually treated as a state estimation problem. *State estimation* is the process of estimating the state

This place could be one of many similar places in the maze.

Figure 19.1   Problems associated with localization.

of a system from measurements. This is a hard problem for the following reasons:

1. *The estimation process is indirect.* One does not usually have the luxury of directly measuring the quantity. Instead, one measures what one can, and estimates the state based on those data ("data" is the plural form of the word "datum"). As we discussed above, a robot may keep track of its wheel rotations, and thus use odometry in order to estimate its location. It is not measuring the location directly; it is estimating it from odometry measurements.

2. *The measurements are noisy.* Of course they are; they involve real-world sensors and real-world properties. By now this should come as no surprise to you, right?

3. *The measurements may not be available all the time.* If they are, the state estimation may be continuous; otherwise, it may be done in batches, when enough data have been collected. For example, odometry is continuous, but other measurements may not be, such as counting mile markers or other types of intermittent landmarks.

State estimation usually requires a model of the relevant quantities in the environment. For odometry, the only model needed is the reference frame in which the distance is measured. However, more complex types of localization require more complex models. For example, consider a robot that has a range sensor (sonar or laser) and a map of the environment. The robot can take a set of range measurements and compare those with the map in order to see where it is most likely to be. This is a tricky process, because several places in the environment may look the same to the particular sensors. Suppose that the robot is in a room that has four empty corners, all of which look the same as the robot is facing them. In order to tell them apart, the robot has to move around and perceive other features (walls, furniture, etc.) so it can see where it is, relative to the map. This type of confusion is very common in robot navigation, whether from corners, corridors, doorways, or other features that often look similar, especially to robot sensors, which are limited and noisy.

As discussed in Chapter 12, maps are just one way to represent the environment, and they come in all sorts of varieties. Instead of having a detailed map, the robot may know about only a few landmarks or targets, in which case it has to compare its sensor readings (i.e., measurements) to the model it has of those landmarks. It may only know their distances from each other, and will have to calculate where it is based on that information. The localization problem varies according to the type of representation used.

If the robot has a topological map (remember that concept from Chapter 12?), localization means it has to be able to uniquely identify which landmark of that map it is in. This is not so hard if the nodes are unique (such as the Statue of Liberty; there is only one in New York City), but is much harder if there are multiple landmarks that look similar (such as a busy street corner with a coffee shop and a newspaper stand; there are plenty of those in New York City). If the robot has a metric map, localization means it has to be able to identify its global Cartesian position (latitude and longitude). Fortunately, that position is sure to be unique.

LOCALIZATION   In summary, *localization* is the process of figuring out where the robot is relative to some model of the environment, using whatever sensor measurements are available. As the robot keeps moving, the estimate of its position drifts and changes, and has to be kept updated through active computation. So knowing where you are (when you are a robot) is not a trivial matter, and being told to "Get lost!" is probably the easiest command to follow.

19.2   Search and Path Planning

As we saw above, path planning involves finding a path from the robot's current location to the destination. This naturally involves the robot knowing its current location (i.e., being localized) and knowing the destination or goal location, both in a common frame of reference. It turns out that path planning is a pretty well understood problem, at least when all of the above information is available to the robot. Given a map of the environment and two special points on it (current and goal locations), the robot just has to "connect the dots."

Typically, there are many possible paths between the start and the goal, and finding them involves searching the map. To make this searching computationally efficient, the map is usually turned into a *graph*, a set of nodes (points) and lines that connect them. Why bother? Because graphs are easy to search using algorithms that have been developed in computer science and artificial intelligence.

A path planner may look for the *optimal* (the very best) path based on some criterion. A path may be optimal based on distance, meaning it is the shortest, it may be optimal based on danger, meaning it is the safest; or it may be optimal based on scenery, meaning it is the prettiest. Admittedly, this last criterion is not usually used in robotics; the point is that there are various criteria for path optimality, depending on the particular robot's task and environment.

Finding the optimal path requires searching all possible paths, because if the planner does not check all possibilities, it may miss the very best one. Therefore, this is a computationally complex (read: potentially slow) process. In fact, path planning is one of those tasks that require robots to perform higher-level thinking or reasoning. This ends up being an important capability that has major impact on how we program robots, as you saw in Chapter 11.

Not all planners look for the optimal path, or even a complete path. Some perform only local path planning, looking at the nearby part of the map, not the whole thing, and therefore speeding things up (but potentially running into problems like closed doors and walls later on). Other planners look for the first path that gets to the goal, rather than the optimal one, again in order to save time. As we talked about at length in Chapter 11 and thereafter, time is very valuable to a robot, and path planning is one important task in a mobile robot's life that can be rather time-consuming. On the other hand, getting to where the robot needs to go is important.

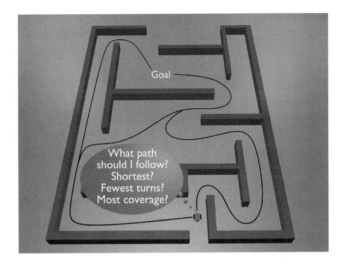

Figure 19.2 Problems associated with path planning.

There is a great deal of work in robotics on various ways to represent the environment, plan a path, and convert that path into a set of movement commands to the robot. Check out some of the relevant books listed at the end of the chapter.

19.3 SLAM

As you can imagine, SLAM is a difficult problem, since it involves having the robot perform two ongoing and related parallel processes: figuring out where it is and constructing a map of the environment. What makes this difficult is the apparent confusion among multiple places that look similar and are therefore ambiguous. This is the *data association problem*, the problem of uniquely associating the sensed data with absolute ground truth.

DATA ASSOCIATION
PROBLEM

If the robot is constructing a topological map, it has to contend with the challenge of uniquely identifying landmarks we described above. If it sees a busy street corner with a coffee shop and a newspaper stand, how does it know exactly which landmark it is seeing? For that matter, even if it is standing next to a large statue holding a torch, how does it know that there aren't two such statues in New York City? It doesn't, until it has the complete map, and by the definition of SLAM, it does not have a map – it is building it as it goes along.

If the robot is constructing a metric map, it has to contend with the errors in odometry and other sensor measurements that make it hard to be sure what global (x,y) position it is actually at. For example, when it goes around a few blocks and corners, can it be sure it has come back where it started? Or is it a block off?

In short, it is easier if the robot can localize when building a map, or if it can have a map in order to localize, but fortunately robotics researchers are working out some nice methods for solving the interesting SLAM problem that forces the robot to do both.

19.4 Coverage

The coverage problem has two basic versions: with a map and without a map.

If the robot has a map of the environment, then the coverage problem becomes one of checking all the navigable spaces of the map until either the object being sought is found or the whole map is covered. (This of course assumes the object does not move; if it does, it is a different and harder problem having to do with pursuit that we'll mention in Chapter 20.) Fortunately, this is clearly a search problem, and therefore there are nice algorithms developed in computer science and AI to solve it, however slowly. For example, there is the British Museum algorithm, an exhaustive brute-force search inspired by the problem of visiting all of the exhibits in that vast museum in London.

HEURISTICS   There are also various search algorithms that use clever *heuristics*, rules of thumb that help guide and (hopefully) speed up the search. For example, if the robot knows that the object is most likely to be in a corner, it will check corners first, and so on.

So if the robot has a map, it uses that map to plan how to cover the space carefully. But what if it does not have a map? In that case it has to move in some systematic fashion and hope to find what it is looking for. This is a hard problem since the environment may have a really complicated shape and make it quite difficult to cover it all without knowing what it looks like. In that case, mapping it first may be a better approach. But that takes time, as we saw.

There are various heuristics for coverage of unknown environments as well. For example, one approach follows continuous boundaries, and another spirals out from a starting point. Still another approach moves the robot randomly, and given enough time and a closed environment, the robot will cover it all. As you can see, which approach and what heuristics to use depend on the robot's sensors, task requirements (how quickly and/or thoroughly it has to get the job done), and whatever it can figure out about the environment.

To Summarize

- Navigation is perhaps the very oldest and most studied problem in mobile robotics.

- Navigation involves a great many subproblems, which include odometry, localization, search and path planning, path optimization, and mapping. Each of these is a separate field of study.

- Today's robots are capable of navigating safely in the environment, but there are still many elements of navigation that can be greatly improved, so the areas above continue to be subjects of research and development.

Food for Thought

- Ants are tremendously effective at dead reckoning, finding their way in the vast open spaces of the desert through odometry. They seem to be much better at it than people, based on some studies that have been done. Of course, no studies have actually pitted a human against an ant in the desert, but comparable experiments find people lost and confused, and ants on a nearly straight path back home.

- A great deal of research has gone into understanding how rats navigate, since they are very good at it (so are ants, as we noted above, but their brains are much harder to study, being so small). Researchers have put rats into regular mazes and water mazes, and even mazes filled with milk. One thing that is very interesting about rats that learn a maze well is that they may run into a wall (if a new wall is stuck in by the experimenter) even if they can see it, seeming to run based on their stored map rather than their sensed data. What type of robot control system would produce such behavior? Rats do it only once, though.

Looking for More?

- The Robotics Primer Workbook exercises for this chapter are found here: http://roboticsprimer.sourceforge.net/workbook/Navigation

- To learn about the *search* problem, check out the following excellent textbooks on computer algorithms: *Introduction to Algorithms* by Corman, Leiserson, and Rivest, and AI: *Artificial Intelligence, A Modern Approach* by Stuart Russell and Peter Norvig.

- To discover how mathematical statistics and probability are used to help robots cope with uncertainty in navigation, brace yourself and ready *Probabilistic Robotics* by Sebastian Thrun, Wolfram Burgard, and Dieter Fox Building. Warning: this is not simple reading and requires a fair amount of mathematics background.

- *Autonomous Mobile Robots* by Roland Siegwart and Illah Nourbakhsh goes into more details about mobile robot kinematics and localization than we did here and is a good resource.

- Here are a couple of textbooks on robot navigation you might want to check out: *Principles of Robot Motion: Theory, Algorithms, and Implementations* by H. Choset, K. M. Lynch, S. Hutchinson, G. Kantor, W. Burgard, L. E. Kavraki and S. Thrun and *Planning Algorithms* by Steven M. LaValle. Note that these are the same as the texts we recommended for manipulation, since the underlying planning problem is the same, it just gets harder in higher dimensions.

20 *Go, Team!*
Group Robotics

What makes a group of robots a team? If you can control one robot, what's different about getting a group of them to do something together? Why is it a hard thing to do well?

Controlling a group of robots is an interesting problem that presents a whole new set of challenges compared with controlling one robot. These include:

1. Inherently dynamic environment

2. Complex local and global interaction

3. Increased uncertainty

4. The need for coordination

5. The need for communication.

Having an environment full of robots creates a very fast-changing, dynamic world. As we already know, the more and the faster the environment changes around the robot, the harder control can be. Multiple robots inherently create a complex, dynamic environment because each moves around and affects the environment and other robots around it. All this naturally leads to increased complexity of the environment, and also more uncertainty for everyone involved. A single robot has to contend with uncertainty in its sensors, effectors, and any knowledge it has. A robot that is part of a group or team also has to deal with uncertainty about the other robot's state (Who is that guy? Where is that guy?), actions (What is he doing?), intentions (What is he going to do?), communication (What did he say? Did he actually say that? Who said that?), and plans (What does he plan to do? Does he have a plan? Will that interfere with my plan?).

In a group situation, a robot has a whole new type of interaction to deal with, one that involves other robots. Just like people, groups of robots must coordinate their behavior effectively in order to get a job done as a team. There are different kinds of "jobs" that teams can do, and different kinds of teams. *Coordination* has to do with arranging things in some kind of order (*ord* and *ordin* mean "order" in Latin). Notice that the definition does not specify how that order is achieved. *Cooperation*, on the other hand, refers to joint action with a mutual benefit. As you will see, some robot teams are coordinated without really being cooperative, while others must cooperate in order to coordinate team performance.

COORDINATION

COOPERATION

Quite a large area of active robotics research is concerned with the challenges of coordinating teams of robots. It is variously called group robotics, social robotics, team robotics, swarm robotics, and multi-robot systems. Let's see what kinds of things it studies.

20.1   Benefits of Teamwork

To begin with, let's consider why we would want to use more than one robot at a time. Here are some potential benefits of using multiple robots:

• **It takes two (or more)**: Some tasks, by their very nature, simply cannot be done by a single robot. Perhaps the most popular example is pushing a box (as in figure 20.1), a prototypical task that has to do with transporting a large, heavy, awkward, or fragile object that cannot be effectively handled by a single robot. Cooperative transportation is useful in construction and habitat building, clearing barriers after a disaster, moving wounded people, and many other real-world applications that require teamwork.

• **Better, faster, cheaper**: Some tasks do not require multiple robots, but can be performed better by a team than by an individual robot. Foraging is the most popular example. *Foraging* is the process of finding and collecting objects (or information) from some specified area. Foraging has been studied a great deal, because it is a prototype for a variety of real-world applications of group robotics, such as locating and disabling land mines, collectively distributing seeds or harvesting crops in agriculture, laying cables in construction, covering an area with sensors in surveillance, and so on.

FORAGING

Figure 20.1 A robot team collectively pushing a box; one robot (behind the box) is the watcher, which steers the team toward the goal, while the two others are the pushers. In this case, it takes three to tango, or at least to push this particular box. (Photo courtesy of Dr. Brian Gerkey)

Foraging can be performed by a single robot, but if the area to be foraged is large compared with the size of the robot, which is usually the case, the job can be performed faster by a robot team. However, determining the right number of robots for such a task is a tricky proposition, since too many will get in each other's way, and too few will not be as efficient as a larger team might be. Determining the right *team* size is just one of the challenges of group robotics. We'll talk about others later in this chapter.

• **Being everywhere at once:** Some tasks require that a large area be monitored so that robots can respond to an emergency event if/wherever it oc-

SENSOR-ACTUATOR NETWORKS

curs. Sensor networks are a popular example. *Sensor-actuator networks* are groups of mobile sensors in the environment that can communicate with each other, usually through wireless radio, and can move about. By having a large number of sensors in a building, an intruder or an emergency can be detected in any area, and nearby robots can be sent to fix the problem. Such networks are also used for *habitat monitoring* – for example, for tracking animals and fish in the wild in order to conserve their environment, for de-

Figure 20.2 Robot dogs playing soccer. (Photo courtesy of Messe Bremen)

tecting and measuring oil spills in the ocean, and for tracking the growth of algae or some contaminant in the water supply. The more such robots there are, the larger the area that can be monitored.

ROBUSTNESS

REDUNDANCY

• **Having nine lives**: Having a team of robots can result in increased robustness in performing a task. *Robustness* in robotics refers to the ability to resist failure. In a team of robots, if one or a small number of robots fails, the others can make up for it, so that the overall team is more robust than any individual alone would be. This type of robustness results from *redundancy*, the repetition of capabilities on the team. Of course, not all teams are redundant: if every member on a team is different, and thereby none can make up for the failures of any others, then the team is not redundant, and is accordingly is not very robust, since the whole team depends on each and every member, and the failure of any one team member can break down the entire team. This type of team organization is usually avoided everywhere, from sports to armies.

20.2 Challenges of Teamwork

The above-listed advantages of robot teams have their price. Here are some disadvantages that create the challenges of multi-robot control:

INTERFERENCE

• **Get out of my way!** *Interference* among robots on a team is the main challenge of group robotics. The more robots there are on a team, the more chance there is that they will interfere with each other. Since the laws of physics just won't allow two or more robots to be in the same place at the same time, there is always the potential for physical interference on a team. Besides this basic type of interference which stems from the robot's physical embodiment, there is also another kind: goal interference, which has to do with robots' goals conflicting. One robot can undo the work of another, intentionally or otherwise, if their goals are conflicting. It turns out that understanding, predicting, and managing interference is one of the great challenges of working with more than one robot at a time.

- **It's my turn to talk!** Many multi-robot tasks require the members of the team to communicate in some way. The most popular means of communication is through wireless radio. A whole field of engineering deals with communications and all the challenges that come with it. As in human communication, there are interference issues here as well. If the robots share a communication channel, they may need to take turns sending messages and receiving them, to avoid "talking over each other", which is the same as collisions in communication-space (rather than in physical space, as in physical interference). Simply put, the more robots there are, the more communication may be useful but also challenging.

- **What's going on?** We have already seen that uncertainty is an inescapable part of a robot's life. The more robots are involved in a task, the more uncertainty there can be. Similarly, the less each individual robot team member knows, the harder it is to get the whole team to agree, in case such agreement is needed (such as in moving to the next part of the task, or recognizing when the task is over; not all tasks require this, as you will see shortly). Some clever multi-robot controllers use multiple members of a robot team to reduce uncertainty, but this does not come for free: it requires clever algorithms and communication among robots.

- **Two for the price of one?** Even with discounts for bulk, more robots are always costlier than fewer, and more robots have more hardware failures and maintenance requirements to worry about as well. Cost is a practical issue that influences real-life decisions in robotics as it does everywhere else.

The challenges of group robotics are what makes programming a team of robots a new and different problem from programming one robot. Before we look into how to program them, let's see what kinds of robot teams there are.

Figure 20.3   A wheeled robot and a six-legged robot cooperatively pushing a box. (Photo courtesy of Dr. Lynne Parker)

20.3   Types of Groups and Teams

How would you program a bunch of robots to play soccer (such as those in figure 20.2)?

Would you program each robot to act is if it were alone – to chase the ball and try to take it, run for the goal and shoot, and run back and protect its own goal or goalie? What would happen if you did? (What would happen if that were how people played soccer?) It would be a complete mess! All robots would run around and into each other, trying to play both offense and defense, and getting into each other's way. Soccer, like most other group activities, requires teamwork and some type of *division of labor* or *role assignment*, giving each member of the team a job so each knows what to do to help the team as a whole without getting in the way of the others.

DIVISION OF LABOR
ROLE ASSIGNMENT

Do all group tasks require division of labor? What about foraging? Could you just have a group of identical robots all doing foraging and staying out of each other's way?

Definitely. This shows us that there are different types of tasks and different types of collective multi-robot systems.

First, teams can be most easily divided into two types, based on the individuals that constitute them:

HOMOGENEOUS
TEAMS

- *Homogeneous teams* are those with identical, and therefore interchangeable, members. Members may be identical in their form (such as all having four wheels and a ring of sonars) and/or in their function (such as all being

Figure 20.4   A highly effective foraging group of merely coexisting robots. (Photo courtesy of Dr. Chris Melhuish)

able to find and pick up objects and send messages to other team members). Homogeneous teams can be coordinated with simple mechanisms, and in some cases require no active, intentional cooperation to achieve effective group behavior (remember emergent flocking from Chapter 18?)

HETEROGENEOUS
TEAMS

- *Heterogeneous teams* are those with different, non-interchangeable members. Members may differ in form (some have four wheels, while others have two wheels and a caster) and/or in function (some play offense by chasing the ball while some play defense by getting in the way of opponents). Heterogeneous teams typically require active cooperation in order to produce coordinated behavior.

Figures 20.1 and 20.3 show examples of real-world robot teams pushing boxes. Both teams are heterogeneous, but one is heterogeneous only in form (different bodies, same function), while the other is heterogeneous in both form and function. Can you tell which is which? The first, a three-member team, features members with different roles: two pushers and a watcher, so it is heterogeneous in both form and function. The second, a two-member team, features two members with different from (wheels vs. legs) but identical function (pushing the box).

The next way of classifying robot teams is based on the type of coordination strategy they use:

Merely coexisting: In this approach, multiple robots work on the same task in a shared environment, but do not communicate or, in some cases, even recognize each other. They merely treat each other as obstacles. Such systems require no algorithms for coordination or communication, but as the number of robots grows and the environment shrinks, interference increases, quickly removing any benefit of having a group. Foraging, construction, and other tasks can be achieved with this approach as long as the team size is carefully designed to suit the task and environment so that interference is minimized.

Let's take a closer look, using the foraging example: We have a group of foraging robots whose job is to look over a field for scattered objects, pick them up, and bring them back to a particular deposit location. While they do that task, they must avoid collisions with obstacles, including other robots. You can see how the more robots are doing the task, the more potential interference there is between them, as long as they are in a bounded, limited amount of space (which is reasonable to expect, since foraging in the vastness of outer space is not currently a major robotics application). In this approach, the robots do not help each other or even recognize each other. Therefore, to make this work, there is a carefully designed balance among the task parameters (including the size of the space and the number of objects being collected), the number of robots, their physical size relative to the environment, their sensor range, and their uncertainty properties. As you can see, this takes some work. Figure 20.4 shows a group of robots that very effectively collected objects into a single pile through merely coexisting. We'll see later how they did it.

Loosely coupled: In this approach, the robots recognize each other as members of a group and may even use simple coordination, such as moving away from each other to make space and minimize interference. However, they do not depend on each other for completing the task, so members of the group can be removed without influencing the behavior of the others. Such teams are robust, but difficult to coordinate to do precise tasks. Foraging, herding, distributed mapping, and related group tasks are well suited for this approach.

Back to our foraging example. Now, instead of treating each other as obstacles, the robots can actually react to each other in more interesting ways. For example, a robot that has found no objects can follow another robot that is carrying an object, in hopes that it will lead it toward more objects. Alternatively, a robot can avoid other robots under the assumption that it should go where others haven't been yet, and so find as yet undiscovered objects.

Figure 20.5 The Nerd Herd, one of the first multi-robot teams, flocking around the lab. (Photo courtesy of the author)

The robots can also flock (as shown in figure 20.5) or form lanes as they are heading to deposit the objects, in order to minimize interference.

Tightly Coupled: In this approach, the robots cooperate on a precise task, usually by using communication, turn-taking, and other means of tight coordination. They depend on each other, which gives the system improved group performance, but also less robustness through redundancy. Removing team members makes the team performance suffer. Playing soccer, moving in formation, and transporting objects are done well through the use of tightly coupled teams.

Returning to foraging yet again: A tightly coupled foraging team would likely use a global plan in which each member of the team goes to a specific area of the environment to explore. This is convenient when a map of the area is available. If no map is given, then the robots may use collaborative SLAM to build one (remember SLAM from Chapter 19?) as they coordinate their exploration and object collection.

20.4 Communication

Let's talk about talking. Members of a team usually (but not always) need to communicate with each other in order to achieve the collective goal(s).

Why would robots communicate?

Here are some good reasons:

1. Improving perception: Robots can sense only so much; by using communication they can know more about the world, without having to sense it directly, by exchanging information with other robots.

2. Synchronizing action: Because robots on a team usually can't perceive all others on the team instantaneously, if they all want to do (or not do, or stop doing) something together at the same time, they need to communicate, or signal, to each other.

3. Enabling coordination and negotiation: Communication is not required for coordinated behavior, as you will see below. However, in some cases and for some tasks, it helps a great deal for robots to be able to cooperate and negotiate in order to get things done right.

The reasons to communicate give us some ideas as to *what* robots might communicate. Consider the options through the example of foraging:

- Nothing: That's what happens in merely coexisting, and it can work very well, as you will soon see.

- Task-related state: The location of the objects, the number of recently seen robots, etc.

- Individual state: ID number, energy level, number of objects collected, etc.

- Environment state: Blocked exits and paths, dangerous conditions, new-found shortcuts, etc.

- Goal(s): Direction to the nearest object, etc.

- Intentions: I'm going this way because I've found objects there before; I'm not going that way because there are too many others there; etc.

And how can robots communicate?

We have already mentioned wireless radio communication, but that's not nearly the whole story. Consider the myriad ways in which people *communicate information*: we gesticulate, we shout to a crowd, we whisper to a friend, we post signs, we email, we leave phone messages, we write letters, cards, papers, and books, and so on. Interestingly, most of those forms of communication can be used by robots. Let's see what options there are for robots who want to communicate:

EXPLICIT COMMUNICATION INTENTIONAL COMMUNICATION

Explicit communication, sometimes also called *intentional communication*, requires an individual to purposefully behave in a way that will convey a message. In robotics that typically involves sending a message over the communication channel, these days wireless radio.

Because wireless communication is ubiquitous (found everywhere), multi-robot communication has gotten a lot easier. Robotics can now rather easily

BROADCAST COMMUNICATION PEER-TO-PEER COMMUNICATION PUBLISH-SUBSCRIBE COMMUNICATION

use *broadcast communication*, sending a message to everyone on the communication channel, or *peer-to-peer communication*, sending a message to a selected recipient.

Publish-subscribe communication is much like using an email list or a news group: a select group of recipients interested in a particular topic signs up for the list, and only those on the list receive messages.

Explicit communication involves a cost because it requires both hardware and software. For any given team task, the designer must think hard about whether communication is needed at all, and if it is, what its range and type (among those listed above) should be, what the information content should be, and what performance level is needed and can be expected from the communication channel. (This last is another reality of life: all communication channels are imperfect, so some messages and parts of messages are lost or corrupted by noise.)

Thus, if the task requires the robots to be able to negotiate one-on-one and in a short amount of time, that presents a very strict set of communication requirements which are fundamentally different from those that may be involved in a task that can be achieved with occasional broadcast messages to all team members.

The forms of communication described so far have all been explicit, used with the purpose of communicating. However, another very powerful and effective type of communication exists:

IMPLICIT COMMUNICATION

Implicit communication involves an individual leaving information in the *environment*, thereby communicating with others without using any explicit communication channels.

STIGMERGY

Stigmergy is the form of communication in which information is conveyed through changing the environment. Ant trails are a perfect example: as ants

PHEROMONE

move about, they leave behind small amounts of *pheromone*, a fragrant hormone that can be detected by others of the same ant species. Ants tend to follow other ants' pheromone trails, and so the more ants go down a certain path, the stronger the pheromone trail becomes and the more ants are recruited. This is a great example of positive feedback, which results from the feedback control we learned about in Chapter 10. In *positive feedback*, the more something happens, the more it "feeds" on itself, and so the more it happens, and so on. Examples of positive feedback behavior include stampedes, lynch mobs, grazing patterns of cattle, termite nests, and of course ant trails. This type of feedback is said to be *amplifying* because it makes some signal or behavior stronger. Positive feedback is the opposite of *negative feedback*, which is *regulatory*, making sure that a system does not go out of control, but rather that it stays close to the set point, just as we discussed in detail in Chapter 10.

So, positive feedback can result not only from explicit but also from implicit communication such as stigmergy, communication through sensing the effects of others in the environment. Robotics researchers have shown how a team of merely coexisting robots that has no way of sensing each other as any different from obstacles can reliably build a barrier using pucks, or collect all pucks within a large area into a single group, purely through stigmergy. Can you figure out how that could be done?

> How can you get a team of robots that can't detect each other but can only tell if they have bumped into something, and that can push a puck, go forward, turn, and back up, to reliably collect all pucks in a single "pile" (the quotes are there because the pile is in 2D, on the flat floor)?

Assume that the environment looks just like that shown in figure 20.4, because those really are snapshots of that very system.

As you have learned already, *the form of the robot (its body) must be well matched to its function (its task)*. The roboticists who designed the puck-collecting robots, Ralph Beckers and Owen Holland, really embraced that principle and used it very ingeniously. They designed a very simple robot, shown in figure 20.6, which had a scoop that served a dual purpose: (1) it collected pucks, because as the robot moved forward, the scoop scooped pucks up and pushed them along; and (2) it detected collisions, because when it was pushed in by the weight of the puck(s), it triggered a 1-bit contact sensor (a simple switch). Using that clever physical robot design, Holland and Beckers then developed a suitably clever controller:

POSITIVE FEEDBACK

AMPLIFYING
NEGATIVE FEEDBACK
REGULATORY

Figure 20.6 A close-up of the cleverly designed puck-collecting robot, part of the swarm that piled up pucks using a single 1-bit contact sensor and no communication. (Photo courtesy of Dr. Chris Melhuish)

```
When hard contact detected
   stop and back up, then turn and go

When soft contact detected
   turn and keep going
```

Amazingly, that was it! The robot's scoop was calibrated (remember calibration from Chapter 8) very carefully not to be too sensitive, so that it could collect quite a few pucks (six to eight) and push them along. Only when it encountered a stronger barrier or heavier weight (more than about six to eight pucks) was its contact switch triggered. Then it stopped, backed up, turned, and moved on, just as its simple minimalist controller told it to do.

Consider what that looks like when you put the robot in a room full of pucks. The robot will move around and, by chance, scoop up pucks. It won't even be able to distinguish pucks from other objects in the environment. Fortunately, there is nothing else in the environment besides pucks, walls, and other robots. We'll get to those things in a bit.

When the robot collects more than eight or so pucks in its scoop, the pucks will be so heavy that its contact sensor will be triggered, and it will stop and back up. And what will that do? It will cause it to leave the pile of pucks there and go on without it! In this way, it will keep making little piles of pucks, and whenever it happens upon a pile of pucks and is pushing another pile, it will drop its new pile next to the old, which will make a larger pile. If you let the robot do this for a while, it will end up with quite a few small piles. If you let it run a while longer, it will produce fewer, larger piles. Eventually, after quite a long long time, it will produce a single large pile. And of course if you put multiple robots in the same area with the same controller, the overall task will be achieved faster.

What happens when the robot runs into the wall?

Well, if the wall is rigid, the robot will stop and back up, but also leave any pucks it has next to the wall, which would not be desirable. The designers cleverly made walls out of flexible fabric, so that the scoop would contact them but the robot would not stop; it would only turn away, keeping the pucks (at least most of the time).

What happens when the robot runs into another robot?

If the contact is soft, it goes on with its pucks, and if the contact is hard, which is the case when robots are heading right into each other, both will back up and move away. And notice this clever side-effect: any pucks they are pushing will end up in a single joined pile.

There are some great videos of this system, which was implemented and validated several times. See the end of this Chapter for pointers to more information.

We can learn a lot from this clever system. First, it is a superb example of effectively using both the robot's physical design and its simple controller to exploit the dynamics of the robots' interaction with each other and the environment in order to accomplish their task. In this system, robots cannot detect each other or communicate with each other. This means they cannot coordinate where the pile of pucks will end up, since that is the result of the system dynamics and is different each time the system runs. On the other hand, it's hard to imagine a simpler and more elegant controller and design for this rather complex task.

The point of the above example is not to say that robots should always use only a single contact sensor and no communication to do collective foraging.

Instead, the point is to show that there are a great many truly innovative and powerful solutions to problems when one thinks broadly and gives up assumptions that may be incorrect, such as "The robots must communicate to get the job done well" or "The robots must recognized each other in order to cooperate." Let's talk about that last one a bit more.

20.4.1 Kin Recognition

Do you think it would be easier for robots in the example above if they could distinguish other robots from pucks and walls?

The answer is yes, if you wanted to write a more complex controller that took advantage of that information. However, that requires more complex sensing and sensory data processing; as you have already seen in Chapters 8 and 9, object recognition is a hard problem. Still, being able to recognize "others like me" is a very useful thing, and is something that most animals do, in multiple ways, through the use of multiple sensors ranging from pheromones (yes, those again), to sound, to vision.

KIN RECOGNITION   In nature, *kin recognition* refers to being able to recognize members of the immediate family, those that share genetic material with the individual. This ability is directly useful in deciding with whom to share food, to whom to signal about predators, and other such potentially "expensive" activities.

In group robotics, kin recognition can refer to something as simple as distinguishing another robot from other objects in the environment, to something as complicated as recognizing one's team members, as in a game of robot soccer. In all cases, kin recognition is a very useful ability, typically worth the sensory and computational overhead it may involve.

Without kin recognition, the types of cooperation that can be achieved are greatly diminished. Kin recognition does not necessarily, or even typically, involve recognizing the identities of all others. Even without identities, sophisticated coordination and cooperation is possible. For example, a team

DOMINANCE HIERARCHY   of robots, just like a group of hermit crabs, can establish a *dominance hierarchy*, more informally known as a *pecking order*. A pecking order helps to give structure and order to a group, so there is less interference.

20.5   Getting a Team to Play Together

So how can we control a group of robots?

There are basically two options: centralized control and distributed control, with some compromises thrown in. Let's take a look.

20.5.1   I'm the Boss: Centralized Control

CENTRALIZED
CONTROL

In *centralized control*, a single, centralized controller takes the information from and/or about all of the robots on the team, thinks about it as long as it takes (and it could take a while if the team is any size above three or so and is doing anything not too trivial in the real world), and then sends commands to all of the robots as to what to do.

No doubt you can readily see the many problems with that idea. It requires a lot of information to come together in a single place; it requires global communication; it is slow and gets slower as the team gets larger; and it is not robust because the centralized controller is a bottleneck in the system: if it fails, the system fails.

Why would we even consider this type of robot team organization?

Because it has one advantage: centralized control allows the team to compute the optimal solution to a given problem it is facing. Putting it more simply: too many cooks can spoil the broth. In an ideal world, if the centralized planner has all the needed correct and complete information and is given enough time to think and make decisions, then the team as a whole can be given perfect instructions.

Does that sound familiar? It should: it's the basic idea of centralized planning which you saw in Chapter 13, so you already know the pros and cons of this approach.

20.5.2   Work It Out as a Team: Distributed Control

DISTRIBUTED
CONTROL

In *distributed control*, there is no single, centralized focus of control; instead, it is spread over multiple or even all members of the team. Typically, each robot uses its own controller to decide what to do.

There are many advantages to this approach: no information needs to be gathered centrally, so there are no bottlenecks and communication can be minimized or sometimes avoided. As a result, distributed control works well with large teams, and does not slow down in performance as the team grows or changes in size.

But as usual, there is no such thing as a free lunch: distributed control brings its own set of challenges. The largest is the issue of coordination: distributed control requires that the desired group-level collective behavior be produced in a decentralized, non-planned fashion from the interactions of the individuals. This means that the job of designing the individual, local behaviors for each robot is more challenging, because it needs to work well with the others in order to produce the desired group behavior.

Given a set of individuals, whether they be bugs or people or robots, it is difficult to predict what they will do together as a group, even if we know what rules or behaviors they execute. This is because what happens locally, between two or three individuals, can have an impact on a large group through positive feedback and other mechanisms of propagation, passing on the effects of actions.

Various sciences, including physics, chemistry, social science, meteorology, and astronomy, have studied such collective behavior of multiple individuals for a long time, and are still at it. The components being studied range from atoms to amino acids to stars and planets to people on the crowded subway. If the individual components in the system being studied are relatively simple, analysis can be somewhat easier, but there are many examples of very simple entities creating very complex collective behaviors (see the end of the chapter for more information). The problem often gets easier if there are very many components, because there exist good statistical tools in mathematics that can be used. In robotics, we are stuck with the hardest of situations, because multi-robot systems have pretty complicated components (even simple robots are not very simple to predict) and not too many of them (until nano robots and smart dust are invented; see Chapter 22 for more on that).

INVERSE PROBLEM Going from local rules to global behavior is a hard problem; it is hard to predict what a lot of robots with particular controllers will do when you let them run. Harder still is the so-called *inverse problem*, which involves going from the global behavior to the local rules. This means it is harder still to figure out what controllers to put on each robot in order to get the group as a whole to do a particular desired behavior (such as puck collection or competitive soccer).

Distributed robot control requires the designer to solve the inverse problem: to figure out what each robot's controller should be, whether they are all the same, or different, i.e., whether the team is homogeneous or heterogeneous, as we saw above, in order to produce the desired group behavior.

20.6   Architectures for Multi-Robot Control

Whether you are controlling a single robot as a centralized controller for a team or a group of robots individually as parts of a team, you are still left with the usual set of control approaches we have already covered: deliberative, reactive, hybrid, and behavior-based control.

Based on all you know, can you figure out which of the control approaches is good for what type of team control?

Your intuition is probably right:

- Deliberative control is well suited for the centralized control approach. The single controller performs the standard SPA loop: gathers the sensory data, uses it all to make a plan for all of the robots, sends the plan to each robot, and each robot executes its part.

- Reactive control is well suited for implementing the distributed control approach. Each robot executes its own controller, and can communicate and cooperate with others as needed. The group-level behavior emerges from the interaction of the individuals; in fact, most of the emergent behaviors seen in robotics result from this type of system: a team of robots controlled using distributed reactive control.

- Hybrid control is also well suited for the centralized control approach, but can be used in a distributed fashion, too. The centralized controller performs the SPA loop, individual robots monitor their sensors, and update the planner with any changes, so that a new plan can be generated when needed. Each robot can run its own hybrid controller, but it needs information from all the others to plan, and synchronizing the plans is hard.

- Behavior-based control is well suited for implementing the distributed control approach. Each robot behaves according to its own, local behavior-based controller; it can learn over time and display adaptive behavior as a result, so the group-level behavior can also be improved and optimized.

20.6.1 Pecking Orders: Hierarchies

It is easiest to imagine how homogeneous robot teams can be controlled with the above control approaches. However, heterogeneous teams with members featuring different abilities and different amounts of control can also be implemented with any of the above.

HIERARCHIES *Hierarchies* are groups or organizations that are ordered by power; the term comes from the Greek word *hierarche* meaning "high priest.' In robot hierarchies, different robots have different amounts of control over others in the team.

As you might guess, two basic kinds of hierarchies exist: fixed (static) and FIXED HIERARCHIES dynamic. *Fixed hierarchies* are determined once and do not change; they are like royal families, where the order of power is determined by heredity. In DYNAMIC contrast, *dynamic hierarchies* are based on some quality that can keep chang-HIERARCHIES ing; they are like dominance hierarchies in some animals that are based on whichever is the largest, strongest, and wins the most battles.

Going back to our different approaches to robot control, here are some options for implementing hierarchies in robot teams:

- Fixed hierarchies can be generated by a planner within a deliberative or hybrid system.

- Dynamic, changing, adaptive hierarchies can be formed by behavior-based systems.

- Reactive distributed multi-robot systems can form hierarchies either by preprogramming or dynamically (e.g., based on size, color, ID number).

Coordinating a group of robots, whether it be a swarm of simple ones or a small group of complex ones, or anything in between, is a major challenge and an active area of robotics research. It is also a highly promising one, since its results are being aimed directly at real-world applications that include autonomous driving and multi-car convoying for safer highway transportation, at sensor network deployment in the ocean for detecting and cleaning up toxic spills, on the ground for fire fighting, for terraforming and habitat construction in space applications, for land-mine detection, farming, reconnaissance and surveillance, and many more. So jump in and join the team!

To Summarize

- Controlling a group of robots is a problem different from controlling a single robot; it involves considerable interaction dynamics and interference.

- Robot groups and teams can be homogeneous or heterogeneous.

- Robot teams can be controlled centrally or in a distributed fashion.

- Communication is not a necessary, but is often a useful, component of group robotics.

- Robot teams can be used in a wide variety of real-world applications.

Food for Thought

- The ways in which robots in a group interfere is not very different from the way people in a group interfere with each other. Some roboticists are interested in studying how people resolve conflict and collaborate, to see if some of those methods can be used to make robots better team players as well. Unlike people, robots don't tend to be self-interested or to show a vast variety of personal differences, unless they are specifically programmed that way.

- It is theoretically impossible to produce totally predictable group behavior in multi-robot systems. In fact, it is a lost cause to attempt to prove or guarantee *precisely* where each robot will be and what it will do after the system is running. Fortunately, that does not mean that multi-robot system behavior is random. Far from it; we can program our robots so that it is possible to characterize, even prove, what the behavior of the group will be. The important fact of life for multi-robot systems is that we can know a great deal about what the group as a whole will do, but we cannot know exactly and precisely what each individual in such a group will do.

- Imagine that the robots in the group can change their behavior over time, by adapting and learning. In Chapter 21 we will study what and how robots can learn. Learning in a team of robots makes coordination even more complex, but it also makes the system more interesting and potentially more robust and useful.

- The ideas of positive and negative feedback are used in common English terminology to refer to giving praise/reward and punishment. Actually, that is an example of loose use of terms; we'll talk about how positive and negative feedback relate to reward and punishment in Chapter 21, which deals with robot learning.

Looking for More?

- The Robotics Primer Workbook exercises for this chapter are found here: http://roboticsprimer.sourceforge.net/workbook/Group_Robotics

- A team of twenty very simple wheeled robots called the Nerd Herd was one of the very first to ever perform group behaviors, such as following, flocking, herding, aggregating, dispersing, and even parking neatly in a row. That was back in the early 1990s at the MIT Mobot Lab, which you already read about in Chapter 5. You can see videos of the Nerd Herd at http://robotics.usc.edu/robotmovies. Here are a couple of papers about how the Nerd Herd was controlled:

  - Matarić, Maja J. (1995), "Designing and Understanding Adaptive Group Behavior", *Adaptive Behavior* 4(1), December, 51-90.

  - Matarić, Maja J. (1995), "From Local Interactions to Collective Intelligence", *The Biology and Technology of Intelligent Autonomous Agents*, L. Steels, ed., NATO ASI Series F, 144, Springer-Verlag, 275-295.

- Here are some of the papers published by the researchers who did the very elegant work on collective sorting we described in this chapter:

  - Beckers, R., Holland, O. E. and Deneubourg J_L (1994). "From Local Actions to Global Tasks: Stigmergy in Collective Robotics," in R. Brooks and P. Maes eds. Artificial Life IV, Cambridge, Mass.: MIT Press, 181-9.

  - Deneubourg, J. L., Goss, S., Franks, N. R., Sendova-Franks, A., Detrain, C., and Chretien, L. (1990). "The Dynamics of Collective Sorting: Robot-like Ants and Ant-like Robots." In Meyer, J-A, and Wilson, S., eds, Simulation of Adaptive Behaviour: from animals to animats, Cambridge, Mass.: MIT Press, 356-65.

- For a look at how complex collective behavior in insects can be, read the impressive book *Ants* by E. O. Wilson.

- Ethology is the science of studying animals in their natural habitat; the term comes from the Greek *ethos* meaning "custom." Textbooks on ethology provide fascinating background about animal social and collective behavior, which has served as inspiration for many roboticists. Ethologists who studied insect behavior and other topics relevant to what we discussed in this chapter include Niko Tinbergen, Konrad Lorenz, David Macfarland, Frans de Waal, and E. O. Wilson. You can find textbooks by each of them.

21 *Things Keep Getting Better*
Learning

LEARNING *Learning*, the ability to acquire new knowledge or skills and improve one's performance, is one of the distinguishing features of intelligence, human or robotic.

How can a robot change its programming and get better at what it does?

There are, it turns out, quite a lot of ways. But first, let's consider what kinds of things a robot can learn.

A robot can learn about itself. Remember that just because a robot can do something, that does not mean it knows it can do it or how well it can do it. So, it is quite useful for a robot to learn how its sensors tend to work and fail (e.g., "If I sense something at location (x,y), where is it really likely to be?"), how accurate its actuators are (e.g., "If I want to go forward by 10 cm, how far am I really likely to go?"), what behavior it tends to perform over time (e.g., "I seem to run into obstacles a lot."), how effective it is at achieving its goal ("I can't seem to achieve my goal for a very long time."), and so on. Such knowledge is not easily pre-programmed, and may vary over time, so it is best learned.

Notice that this gives us a hint about what is worth learning. But we'll talk about that a bit later in this chapter.

Besides learning about itself, **a robot can learn about its environment**. The most popular example is learning maps, which we talked about in Chapter 19. Even without learning a map, a robot can learn paths to a goal, such as learning a way through a maze, as a sequence of moves. The robot can also learn where unsafe areas of the environment are, and much more, based on what about its environment matters for performing its task. By the way, that is another hint about what is worth learning.

As we saw in Chapter 20, for some robots the environment includes other robots. In that case, **a robot can learn about other robots**, things such as how many of them there are, what kinds, how they tend to behave, how well the team works together, and what works well for the team and what does not.

Let's return to the basics: why should a robot bother to learn? Here are some good reasons to consider:

- Learning can enable the robot to perform its task better. This is important because no robot controller is perfect, nor is any robot programmer.

- Learning can help the robot adapt to changes in its environment and/or its task. This is hard to pre-program if changes cannot be anticipated by the programmer.

- Learning can simplify the programming work for the designer of the robot's controller. Some things are just too tedious or too difficult to program by hand, but can be learned by the robot itself, as you will see shortly.

Now that we have the what and why of robot learning covered, we can turn to the real question: how do we get a robot to learn?

21.1   Reinforcement Learning

REINFORCEMENT
LEARNING

Reinforcement learning is a popular approach for getting robots to learn based on feedback received from the environment. The basic idea is inspired by learning in nature, the way animals (including people) learn. Reinforcement learning involves trying different things and seeing what happens; if good things happen, we tend to do the behavior again, and if bad things happen, we tend to avoid it.

This basic process turns out to be a remarkably versatile tool for learning. It allows robots to learn what to do and not to do in various situations. Consider a typical reactive controller that tells the robot how to react under different sensory inputs. By using reinforcement learning, you can have the robot learn such a controller instead of programming it in. Well, close to it, anyway.

Consider the robot's input space (remember what that is from Chapter 3) as the set of all possible situations. Now consider the set of all of its possible actions. If the robot knows what action to perform in each state, it will have a complete controller. But suppose we would prefer not to pre-program the controller, but to have the robot learn it on its own.

How could a robot learn to match the right states with the right actions?

By trial and error! In reinforcement learning, the robot tries different actions (in fact, all of the actions it has at its disposal) in all states it gets into. It keeps track of what happens each time in some type of representation (usually some type of a table or matrix). In order to learn what action is best in each state, it has to try everything; in machine learning (for robots, software agents, and other kinds of programs), the process of trying all possible state-

EXPLORATION action combinations is called *exploration*. Exploration is important, because until the robot has tried all possible actions in all states, it can't be sure that it has found the best action for each state.

Once the robot has tried all combinations of states and actions, it knows what is good and what is bad to do relative to its task and goals. Wouldn't it be great if it were done now? Unfortunately, it's not. Here is why:

- Things are not what they seem: If there is any error in the robot's sensing of the current state or in its execution of its actions, the result of trying any state-action combination may be incorrectly learned. Uncertainty strikes again.

- Things may change: If the environment or task changes while the robot is learning, what has it learned may be invalid.

Both of those reasons make it important for the robot to keep checking what it has learned, not only in reinforcement learning but in any learning approach. In reinforcement learning in particular, the robot has to keep *exploring* instead of always *exploiting* what it has learned. In machine learning,

EXPLOITATION the process of using what has been learned is called *exploitation*. We tend to think of exploitation as some kind of unfair use, but it really means using something to its fullest advantage. If a robot purely exploits what it has learned without exploring further, it will not be able to adapt to changes or continue to learn. The trade-off between constantly learning (at the cost of doing things less than perfectly) and using what is known to work well

EXPLORATION VS. (at the cost of missing out on further improvements) is called *exploration vs.*
EXPLOITATION *exploitation*.

Ok, we know that the robot tries different things and remembers how they turn out. But how does it do that, and exactly what does it remember?

Imagine that the robot has a table, like the one shown in figure 21.1, that has all possible states as the rows, and all possible actions as the columns. Initially this table is empty or all of its entries are the same, say 0. After the robot

Left Bump	Right Bump	Action	Q-Value
on	on	on	0.0
on	on	off	1.0
on	on	on	0.0
on	on	off	0.0
on	off	on	0.1
on	off	off	0.4
on	off	on	0.0
on	off	off	0.5
off	on	on	0.1
off	on	off	0.4
off	on	on	0.5
off	on	off	0.0
off	off	on	1.0
off	off	off	0.0
off	off	on	0.0
off	off	off	0.0

Figure 21.1   A policy table for a simple learning robot.

tries a particular state-action combination and sees what happens, it updates that entry in the table. An intuitive way to think about this is to imagine that the value of a state-action entry in the table grows if good things happen and shrinks if bad things happen. To keep this mathematically correct, the values in the table need to be normalized and updated in a consistent way. Entries can be updated in a variety of different ways, depending on what reinforcement learning algorithm the robot is using; two popular ones are temporal differencing and Q-learning. There are numerous papers and textbooks written about robot learning that describe these algorithms, so we won't spend much time on them here; but you can easily look them up. The most important thing to realize is that the robot uses reinforcement learning to learn state-action combinations. The complete state-action table is called

CONTROL POLICY   a *control policy*; these are reactive rules that program a robot for a particular goal.

Every policy is specific to a particular goal, just as any controller is specific to a particular task. If the goal/task changes, so must the policy/controller.

VALUE FUNCTION   As an alternative to a policy, the robot could learn the *value function*, the value of being in each state relative to the goal. In practice, this is the same as learning a control policy, so we won't discuss it further.

For reinforcement learning to work on a robot, the robot has to be able to evaluate how well a state-action pairing went, what the outcome was of a trial. For example, if the robot is learning how to walk, it has to be able to measure if it has moved forward after trying some state-action combination.

POSITIVE FEEDBACK
REWARD
NEGATIVE FEEDBACK
PUNISHMENT

If it has, that state-action combination receives *positive feedback*, more simply called *reward*. Conversely, if the robot moved backward, it will receive *negative feedback*, more simply called *punishment*. (It is important not to confuse these notions of feedback in reinforcement learning with those we discussed in Chapter 10; as they are not the same.)

Notice that feedback can come in different ways, and usually simple good or bad is not enough. In the example above, the robot will learn about going forward and backward, but it will not learn what to do if the outcome of a state-action pair leaves it stopped.

This is just a simple example that illustrates that setting up the right kind of feedback mechanism is one of the major challenges of reinforcement learning.

> *How does the robot know exactly how much to reward or punish a particular outcome?*

This turns out to be really tricky.

Reinforcement learning works quite well under some specific conditions:

- The problem is not too huge: If there are so many states that it would practically take the robot days to try them all and learn, then the approach may not be desirable or practical; it all depends on how hard it is to program the system by hand in comparison.

- The outcome is clear: The result of tested state-action combinations are known soon after the trial and can be sensed by the robot.

Let's take another example. Consider a robot learning a maze. Taking the right turn early on can make all the difference in eventually finding the way out, but at the time the robot takes the turn, it will not know whether that is a good thing to do or not. In that case, the robot has to remember the sequence of state-action pairs it took, and then send back the reward or punishment to each of them in order to learn. With uncertainty, this can get

TEMPORAL CREDIT
ASSIGNMENT

quite complicated and messy, but not impossible. The general problem of assigning the credit or blame to actions taken over time is called *temporal credit assignment*.

The reinforcement learning problem can also be made more difficult in a multi-robot environment, where it is not always easy to tell which robot's action brought about a good (or bad) outcome. A robot may do an action in a state that does nothing, but if another robot happens to do just the right next thing next to it, either may assume the desirable outcome is due to itself and not the other. This problem is called *spatial credit assignment*.

SPATIAL CREDIT
ASSIGNMENT

Reinforcement learning has been used with great success in non-robotics applications; for example, the world backgammon champion was a computer program that learned the game by playing against itself and using reinforcement learning. Of course in robotics, learning is much messier due to uncertainty, dynamically changing environments, and hidden/partially observable state (remember Chapter 3). Nevertheless, robots have successfully used variations of reinforcement learning to learn to forage, navigate mazes, juggle, pass and score in soccer, and many other activities.

21.2   Supervised Learning

UNSUPERVISED
LEARNING

Reinforcement learning is an example of *unsupervised learning*, meaning that there is no external supervisor or teacher who tells the robot what to do. Instead, the robot learns from the outcomes of its own actions. The natural opposite of unsupervised learning is *supervised learning*, which involves an external teacher who provides the answer or at least tells the robot what it did wrong.

SUPERVISED LEARNING

NEURAL NETWORK
LEARNING

CONNECTIONIST
LEARNING

Neural network learning is a very popular approach for supervised machine learning for robots and other non-embodied machines. A wide variety of neural network learning algorithms exists, all collectively called *connectionist learning*, because typically what is being learned is the weights on the various connections in the networks (how strongly nodes are connected), or in some cases the structure of the network itself (what is connected to what; how many nodes there are). There exists a growing field of *statistical neural networks* in which the learning is done with very mathematically formal techniques from statistics and probability. In fact, the use of statistics and probability in robot control and learning is widespread; see the end of the chapter for pointers to more literature on the topic.

STATISTICAL NEURAL
NETWORKS

Reinforcement learning and neural networks learning are both only distantly related to the way learning works in nature. They are not meant to be careful models of natural systems, but practical algorithms for helping machines improve behavior. So don't get confused by any claims to the contrary; neural networks are not like the brain (brains are much, much more complex) and reinforcement learning is not like training circus animals (circus animals, at least for now, are much smarter).

In unsupervised learning, the robot gets a bit of reward or punishment some of the time. In contrast, in supervised learning, the robot gets the full answer in the form of the magnitude and direction of the error. Consider the case of ALVIN, the autonomous driving robot van which learned to steer on the roads around Pittsburgh, Pennsylvania, through the use of a neural network. Each time it turned the steering wheel, the supervised learning algorithm told it just how it should have done it precisely right. It then computed the *error*, the difference between what it did and what it should have done (just as we defined it in Chapter 10), and used that difference to update its network (the weights on the links). By doing this type of training some hundreds of thousands of times, mostly in simulation (using real images of the road but not actually driving on the physical road), the robot learned to steer quite well.

ERROR

You can see how supervised learning provides much more feedback to the robot than unsupervised learning does. Both require a lot of training trials, and each is suitable for different types of learning problems. Learning robot inverse kinematics and dynamics (remember Chapter 6?) lends itself to supervised learning, where the right answer is readily available and the error can easily be computed. As an example, consider a robot that tries to reach and point at a particular object; the distance from that object and where the robot actually reached and pointed provides the error, and allows the robot to learn by reaching over and over again. Not surprisingly, a great many robots have learned their inverse kinematics in just this way.

As another example, consider maze learning: the robot has no map of the maze (if it did, then it would not be maze learning, just map searching and path planning) and has to learn the best path to the cheese. It does this by trying different paths; if it runs into a dead end in the maze, it learns that was a bad place to go, but it cannot compute the error from the right thing to do, because it does not know the right thing to do (that would require a map of the maze, which it does not have). Therefore, it must use reinforcement learning and discover the path to the cheese that way.

21.3   Learning by Imitation/From Demonstration

Some types of learning are best done by being shown by a teacher. Consider how babies learn: they do a great deal of both supervised and unsupervised learning as described above (reaching, throwing, putting in the mouth and chewing, babbling, crawling, and so on), but they also learn a huge amount from friendly humans.

Learning from demonstration and by imitation is incredibly powerful, because it frees the learner from having to do things by trial and error, and from making large mistakes. The learner, human or robot, gets the perfect example of what to do, and merely needs to repeat what it sees.

If only it were that simple. As it turns out, it is not. In fact, very few species on Earth are capable of learning this way; only humans, dolphins, and chimpanzees are believed to be able to learn arbitrary new skills by seeing them performed by others. The rest of the animals either can't do it at all, or can learn just a few things this way (monkeys and parrots fall into this category). And although many pet owners would disagree, and claim that their particular pet can learn anything by adoringly watching the owner, science so far tells us otherwise.

Imitation is very powerful yet very hard. Fortunately for robots, we can try to understand why it is hard, yet still make it possible and useful. So why is it hard?

In order for a robot to learn by demonstration, it must be able to:

- Pay attention to a demonstration. (Something buzzed, but I can't look away! Do I watch the legs or the arms? Oops, I moved and missed something.)

- Separate what is relevant to the task being taught from all irrelevant information. (Does the color of the shoes matter? What about the specific path being traced?)

- Match the observed behavior to its own behaviors and effectors, taking care about reference frames. (The teacher's right is my left and vice versa. The teacher has two arms but I have two wheels, so what do I do? The teacher is more flexible than I am, my joint limits do not allow me to do what I saw.)

- Adjust the parameters so the imitation makes sense and looks good. (The teacher did this slowly but I can do it faster. Will it look as good and get the job done? Do I reach lower since I'm shorter?)

Figure 21.2 A robot learning by following a teacher and being "put through" the task. (Photo courtesy of Dr. Monica Nicolescu)

- Recognize and achieve the goals of the performed behavior. (Was that a wave or a reach to a particular position in space? Was the object moved on purpose or as a side-effect?)

The above list is not easy for people (just imagine or recall learning a new complicated dance), much less for robots, for all the reasons of uncertainty, limited perception, and constrained actuation we have discussed already in several chapters.

While it may not be easy, learning by demonstration/from imitation is a growing area of robot learning, because it not only promises to make robot programming easier, but it creates the opportunity for people and robots to interact in a playful way. This is important for the rapidly growing field of *human-robot interaction*, which we'll talk more about in the next chapter.

HUMAN-ROBOT INTERACTION

Learning by imitation involves careful decisions about internal representation. In one implemented system, shown in figure 21.2, a mobile robot learns a new task by following the teacher around and doing whatever she does. In the process, the robot experiences a set of very similar inputs to those of the teacher, and can observe itself performing a set of behavior, for example: "Go along the corridor, then grab the red box, and carry it to the next room, and drop it in the corner." By "mapping" its observations to its own behaviors,

Figure 21.3   A robot teaching another robot the same task by "putting through." (Photo courtesy of Dr. Monica Nicolescu)

the robot learns new tasks. For example, the robot was able to learn tasks such as fetching and delivering objects and navigating around an obstacle course, and could then teach those new tasks to another robot, through the same process of demonstration and imitation, as shown in figure 21.3. This type of approach finds inspiration in theories from neuroscience, and works well with modular behavior-based systems.

Putting Through   The idea of having the learner experience the task directly is called *putting through*. This is a very powerful way to learn, and is often used for teaching people skills, such as how to swing in golf or tennis. It helps to try something out and know how it "feels" when done correctly. The same idea can be used in teaching robots. For this to work, the robot needs to have some internal representation of its own behaviors or actions, so it can remember what it experienced and how it can generate that type of behavior again. It can then teach others, too, as shown in figure 21.3.

In another example of robot learning by demonstration, a humanoid robot, shown in figure 21.4, learned to balance a pole on its "finger" by watching a human do it. Unlike humans, the robot could learn the task after only a few trials, over a few minutes, and then perform it perfectly thereafter. How is that possible, and why can't people do that? A part of the reason has to

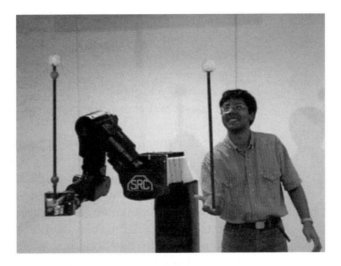

Figure 21.4 A humanoid robot learning to balance a pole by watching a human teacher. (Photo courtesy of Dr. Sethu Vijayakumar)

do with the robot's ability to have a model of its own motor control. In this example, the robot knew exactly how its arms worked, and thus could perfect its pole-balancing by tuning a few parameters after a small number of trials with the pole. Sometimes robots do have an unfair advantage, in this case because the robot had direct computational access to its inverse kinematics and dynamics, while people do not. We have to learn by trial and error, while robots can learn by quickly building a model of what works for their particular explicitly modeled system parameters.

If you remember what we learned in Chapter 9 you may be wondering how the robot could track the pole accurately by using vision while it was moving around, since that is not a simple problem. In fact, the robot watched only the two colored balls on the pole, and the way they moved relative to each other. It had a model of pole-balancing which involved the top ball being above the bottom one, and so it learned how to keep it in that position. What it really learned was to balance a pole (which was attached to those balls), which happens to be a well-studied inverted pendulum problem from control theory (remember that from Chapter 10?).

21.4 Learning and Forgetting

Because robot learning and learning in general are such fascinating and challenging problems, various methods for learning have been developed, and have been tried on robots with different levels of success. Besides the approaches described above, there are memory-based learning, evolutionary learning, case-based learning, statistical learning, and many more. In fact, there are too many to put in one chapter, but hopefully you are curious and will look them up and learn more about them.

Something you must not forget when considering learning is forgetting! Forgetting seems like a bad thing, at least when it comes to human memory, assuming that all information we have ever acquired is worth keeping. However, that is not actually the case, and it is impossible with limited memory. FORGETTING In machine learning, therefore, purposeful and careful *forgetting* is useful, because it allows for discarding outdated previously learned information in favor of newer, more current information. Forgetting is necessary for two reasons:

1. Making room for new information in finite memory space

2. Replacing old information that is no longer correct.

Computer memory these days is cheap, but it's not infinite, so eventually some stuff has to be thrown out to make room for new stuff. Even if memory were to become, in effect, infinite, and robots could store all they ever learned, we would still have some challenges. First, there would have to be methods for very quickly searching all that stored information in order to get to something specific. This is a well-studied problem in computer science; there are ways to organize information so things can be looked up efficiently, but it takes work. More important, given a lot of information on a given topic or task, some of which may be contradictory (because what the robot learns over time naturally changes), how does the robot decide what information/knowledge to use? A simple answer is to use the latest, newest information, but that may not always work, since just because something was learned or experienced recently, that does not mean it is the best approach or solution. For example, the most recent way you saw from home to the store may have been circuitous and slow; you should not reuse that way but instead reuse the path you took last week, which was much shorter.

Knowing something or having some previously learned skill can interfere with learning something new. Rats that learn to run a particular maze have a hard time adapting if a path to the cheese is unexpectedly blocked; they run into the new wall even though they see it, because they believe their learned internal map of the maze. Eventually they relearn, but it takes a while. Similarly, people who have learned to play tennis take a bit to adjust to playing squash, because in the former you have to hold your wrist rigid while in the latter you have to bend it a lot to swing the (smaller and lighter) racquet. Finally, sometimes people absolutely refuse to accept new information or knowledge because it conflicts with something they already know and prefer to believe; we call this denial. These are just a few illustrative examples, ranging from rodents to people, of how learning, unlearning, and forgetting need to work together and can cause problems when they do not.

LIFELONG LEARNING   *Lifelong learning*, the idea of a robot learning continually and constantly improving, is something roboticists dream about but do not yet know how to do. This is a great challenge for many reasons. Aside from learning itself being challenging, for all the reasons outlined so far in this chapter, lifelong learning is also hard because so far we have different approaches to learning different things (from maps to juggling to delivery), as you saw above. Since a robot needs to learn (and forget) a variety of things in its lifetime, such multiple learning methods have to be combined into a single system in some principled and efficient way.

To summarize

- Robot learning is harder than other types of learning because it involves uncertainty in sensing and action, and large and often continuous sensory input spaces and effector action output spaces.

- The amount and type of information (feedback, reward, punishment, error) available to the learning robot determine what type of learning methods are possible for a particular learning problem.

- Various robot learning methods have been studied and continue to be pursued in research, including reinforcement learning, supervised vs. unsupervised learning, learning by imitation/demonstration, and evolutionary learning, to name a few.

- A robot can use multiple learning methods and learn a variety of things at the same time.

- Robot learning is in its infancy, with great new discoveries yet to be made.

Food for Thought

Some people find the idea of robots learning frightening, because they are uncomfortable with the notion of robot behavior being anything but completely predictable. Does adding learning to the robot's abilities make it more unpredictable? If so, how can this be helped? If not, how can people's worries be put to rest?

Looking for More?

- The Robotics Primer Workbook exercises for this chapter are found here: http://roboticsprimer.sourceforge.net/workbook/Learning

- Tom Mitchell's *Machine Learning* is a popular and comprehensive textbook spanning the wide variety of learning approaches.

- Reinforcement Learning: An Introduction by Rich Sutton and Andy Barto is an excellent introduction to this popular learning approach.

- *Robot Learning* edited by Jonathan Connell and Sridhar Mahadevan is a nice collection of real robot learning approaches and results.

- *Probabilistic Robotics* by Sebastian Thrun, Wolfram Burgard, and Dieter Fox Building, the same text we recommended in Chapter 19, covers the uses of statistical and probabilistic approaches in robotics for applications beyond navigation. The same warning applies: this is not simple reading and requires a fair amount of mathematics background.

22 *Where To Next?*
The Future of Robotics

Right now, as you are reading this book, we are at a particularly interesting time in the history of robotics.

The great thing about the above statement is that it seems true no matter when you read the book. Roboticists have been continually excited about the future of their field. Most will admit that they have predicted major breakthroughs at various times, and many such breakthroughs have taken place.

However, the start of the twenty first century is an especially pivotal time for the field. Here is why:

- *Sensors, effectors, and bodies are becoming very sophisticated.* We are creating the most complex robots yet, with bodies that mimic biological form and attempt to model biological function.

- *Computers are faster and cheaper than ever before.* This means robot brains can be more sophisticated than ever, allowing them to think and act efficiently in the real world.

- *Wireless communication is everywhere.* This means robots can communicate with other computers in the environment, so they can be better informed, and therefore smarter.

Other advances have played a role as well. Importantly, as you recall from Chapter 2, robotics got its name from menial labor and has historically done its share of factory work. But now robots are starting to enter our daily lives, and that is going to change everything.

The most pervasive use of robotics until recently was in car factories, as shown in figure 22.1. Robots were perfected to be so effective that they are

Figure 22.1   Robot arms assembling a car. (Photo courtesy of KUKA Schweissanlagen GmbH in Augsburg, Germany)

Figure 22.2   An assembly robot involved in gene sequencing. (Photo courtesy Andre Nantel)

Figure 22.3 Sony's Aibo robot dogs. (Photo courtesy of David Feil-Seifer)

not considered proper robots any more, but fit into the more general category of *factory automation*. At the end of the twentieth century, robots did some finer work of the assembly type: they were instrumental in helping people sequence the human genome. As in car assembly, the brain work was done by people, but the robots did the hard, precise, and repetitive work, as shown in figure 22.2. As a result, the field of genetics took a tremendous leap, and the process is being applied to genome sequencing of numerous other species, toward doing better science and developing better methods for improving human health.

You probably had no idea that robots were involved in sequencing the human genome, since they were far from the spotlight. Hidden in factories and labs, the robots are not getting much exposure. A much more appealing robot that got a great deal more exposure was the Sony Aibo robot dog, shown in figure 22.3. The Aibo went through several generations, in which its overall shape was refined, but in all cases it was universally deemed "cute" and was used in a variety of research and educational projects. Unfortunately, while Aibo was cute, it was neither useful nor affordable, and thus not very appealing to a broad audience.

In the meantime, perhaps the most popular question for roboticists over the decades has been "Can it clean my house?" The answer, finally, is "Yes, it

Figure 22.4   An autonomous vacuum cleaner by iRobot Inc. (Photo courtesy of iRobot Inc.)

can!" Several robotic vacuum cleaners were developed in the last decade of the twentieth century, but one in particular, the small, simple, and affordable Roomba by iRobot, shown in figure 22.4, filled the special niche of an affordable and sufficiently useful novelty high-tech product. Over two million such robots have been sold to households in the United States and beyond, and a wet version that mops the floor, called Scooba, has also been released.

The Roomba is a very important robot, albeit a simple and modest one. Because it is so simple and affordable, the Roomba has the chance to become a part of people's everyday lives (or at least as often as people clean their floors). In this way, numerous people can finally begin to interact with robotics technology in an approachable setting. And as a result, we can hope that many young people will be inspired to take the Roomba and aim to develop the next highly successful robot that will go into millions of homes.

As robotics is poised on the thresholds of our homes, let us consider some of the exciting directions the field of robotics is taking that will shape the future not only of the field but also of our world.

JPL-25888AC

Figure 22.5 NASA's Sojourner, the first rover to roam on Mars. (Photo courtesy of NASA)

22.1 Space Robotics

Working in harsh environments has always been one of the most natural uses for robots. Space is the ultimate frontier, and one where robots have already made significant contributions. NASA's well-known mission to Mars dropped off a very large and well-bundled bouncing "ball" on the surface of the planet. The ball, when it stopped bouncing and rolling, and opened up, revealed the Pathfinder lander. Shortly thereafter, a hatch on the lander opened and a rover drove out. That amazing little rover was Sojourner, shown in figure 22.5. On July 4, 1997, it was the first to drive autonomously around the surface of Mars, collect images, and send them back to Earth. Sojourner's name means "traveler" and was chosen from 3500 entries submitted as part of a yearlong worldwide student competition. Valerie Ambrose, of Bridgeport, Connecticut, who was twelve at the time, submitted the winning essay about Sojourner Truth, an African-American reformist from the Civil War era whose mission was to "travel up and down the land" advocating the rights of all people to be free and the rights of women to participate fully in society. It's good to name good robots after good people.

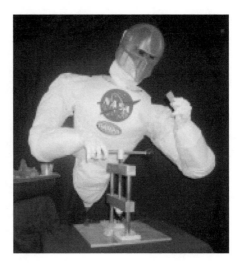

Figure 22.6   NASA's Robonaut, the first space robot with a humanoid torso. (Photo courtesy of NASA)

Since Sojourner, NASA and other space agencies around the world are considering terraforming first the moon, and eventually Mars, with a view to potential human colonization. What used to be science fiction may become the future, and robots will be necessary to make it possible. Currently robots such as NASA's Robonaut, shown in figure 22.6, are being developed to be capable of working side by side with astronauts, as well as completely autonomously, in order to do tasks that range from fixing problems with spacecraft to doing exploration and sample return from planets, moons, and even asteroids, to constructing habitats for human use.

22.2   Surgical Robotics

Let's jump from very large and distant spaces to very small and up-close-and-personal ones, our bodies. Surgical robotics is one of the better-established uses of robots in challenging real-world environments. As briefly mentioned in Chapter 6, robots have been involved in hip surgery and even brain surgery for some time. They are excellent replacements for human workers in those domains because precision is of critical importance (Can you walk without part of a hip or think without part of your brain? Would you want

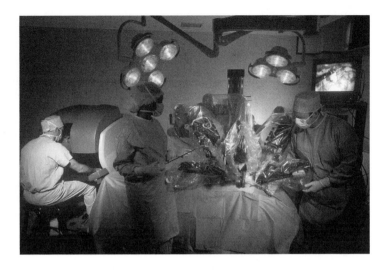

Figure 22.7 A surgical robot at work. (Photo ©[2007] Intuitive Surgical, Inc.)

to?), but intelligence is not. In fact, these machines are not intelligent and autonomous at all, but are remotely controlled by highly skilled surgeons, who not only are well practiced in surgery but also are well trained in using robots to do surgery, which is typically called *robot-assisted surgery*. The surgery is merely "assisted," in that the human surgeon does all the perceiving (looking through the special glasses into what the camera on the robot sees), thinking (deciding what actions to take), and doing (taking the actions with the remote controls). The robot is a teleoperated manipulator, with special endeffectors ideally suited for the particular surgical task. This process is so well worked out that in some cases the surgeon need not be in the same room or even state as the patient; surgery has been performed across continents, with patients in the United States and surgeons in France, and vice versa. This sounds like a great idea for providing surgical aid to people in underdeveloped countries or on the battlefield, until one considers that such robots currently cost about a million dollars. (Perhaps there will be a good sale soon.)

ROBOT-ASSISTED SURGERY

It is easy to imagine the many new directions that surgical robotics research will pursue. Perception is the first challenge; surgeons would like the ability to "feel" the area in which the robot is operating, the way they would if they were doing the work by hand (so to speak). This involves complex

HAPTICS   research challenges in *haptics*, the study of the sense of touch. Interestingly, the term comes from the Greek *haptikos* meaning "being able to grasp or perceive." Take a second to envision the environment of a surgical robot: there are no simple rigid or flat surfaces, and the lighting is less than ideal. Finally, today there are no detailed models of the environment available in advance, though they could be made before surgery with sophisticated imaging equipment, if it is available.

Surgeons are far from ready to give up control to a robot (or anyone?) yet, but if we are to contemplate robots that can perform any of this work on their own, research will have to address real-time decision-making and control within the human body. If that is not scary enough, consider that NANOROBOTICS there is also ongoing research into *nanorobotics*, truly tiny machines that can float freely in our bloodstreams and hopefully do some useful work such as scraping plaque from blood vessels, detecting danger areas, and so on. Needless to say, this research is still in its infancy, since manufacturing a tiny thing requires another tiny thing, and we do not yet have such tiny things SELF-ASSEMBLY available. To counter this problem, some researchers are looking into *self-assembly*, which is exactly what you would guess: the ability for robots to assemble themselves and each other, especially at small scales. That is still mostly in the zone of science fiction, so let's turn to some large and more current directions.

22.3   Self-Reconfigurable Robotics

In theory at least, a swarm of tiny robots can get together to create any shape. In reality, robots are constructed with rigid parts and do not shape-shift eas-RECONFIGURABLE ily. But an area of robotics research called *reconfigurable robotics* has been de-ROBOTICS signing robots that have modules or components that can come together in various ways, creating not only different shapes but differently shaped robot bodies that can move around. Such robots have been demonstrated to shape-shift from being like six-legged insects to being like slithering snakes. We say "like" because six-legged insect robots and snakelike robots have been built before, and were much more agile and elegant than these reconfigurable robots. However, they could not shape-shift.

The mechanics of physical reconfiguration are complex, involving wires and connectors and physics, among other issues to consider and overcome. Individual modules require a certain amount of intelligence to know their state and role at various times and in various configurations. In some cases,

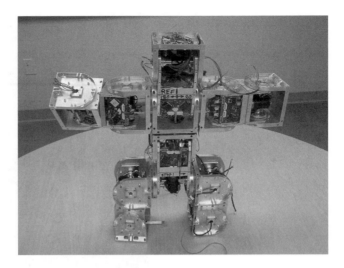

Figure 22.8 A reconfigurable robot striking (just one) pose. (Photo courtesy of Dr. Wei-Min Shen)

the modules are all the same (homogeneous), while in others, they are not (heterogeneous), which makes the coordination problem in some ways related to multi-robot control we learned about in Chapter 20. Reconfigurable robots that can autonomously decide when and how to change their shapes are called self-reconfigurable. Such systems are under active research for navigation in hard-to-reach places, and beyond.

22.4 Humanoid Robotics

Some bodies change shape, and others are so complicated that one shape is plenty hard to manage. Humanoid robots fit into that category. As we learned in Chapter 4, parts of the human body have numerous degrees of freedom (DOF), and control of high-DOF systems is very difficult, especially in the light of unavoidable uncertainty found in all robotics. Humanoid control brings together nearly all challenges we have discussed so far: all aspects of navigation and all aspects of manipulation, along with balance and, as you will see shortly, the intricacies of human-robot interaction.

There is a reason why it takes people so long to learn to walk and talk and be productive, in spite of our large brains; it is very, very hard. Human be-

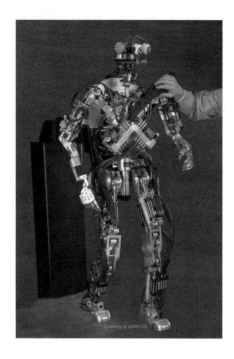

Figure 22.9   Sarcos humanoid robot. (Photo courtest of Sarcos Inc.)

ings take numerous days, months, and years to acquire skills and knowledge, while robots typically do not have that luxury. Through the development of sophisticated humanoid robots with complex biomimetic sensors and actuators, the field of robotics is finally having the opportunity to study humanoid control, and is gaining even more respect for the biological counterpart in the process.

22.5   Social Robotics and Human-Robot Interaction

Robotics is moving away from the factory and laboratory, and ever closer to human everyday environments and uses, demanding the development of robots capable of naturally interacting socially with people. The rather new

HUMAN-ROBOT INTERACTION (HRI)

field of *human-robot interaction (HRI)* is faced with a slew of challenges which include perceiving and understanding human behavior in real time (Who is that talking to me? What is she saying? Is she happy or sad or giddy or mad? Is she getting closer or moving away?), responding in real-time (What should I say? What should I do?), and doing so in a socially appropriate and natural way that engages the human participant.

Humans are naturally social creatures, and robots that interact with us will need to be appropriately social as well if we are to accept them into our lives. When you think about social interaction, you probably immediately think about two types of signals: facial expressions and language. Based on what you know about perception in robotics, you may be wondering how robots will ever become social, given how hard it is for a moving robot to find, much less read, a moving face in real time (recall Chapter 9), or how hard it is for a moving target to hear and understand speech from a moving source in an environment with ambient noise (recall Chapter 8).

Fortunately, faces and language are not the only ways of interacting socially. Perception for human-robot interaction must be broad enough to include *human information processing*, which consists of a variety of complex signals. Consider speech, for example: it contains important nuances in its rate, volume level, pitch, and other indicators of personality and feeling. All that is extremely important and useful even before the robot (or person) starts to understand what words are being said, namely before language processing. Another important social indicator is the social use of space, called *proxemics*: how close one stands, along with body posture and movement (amount, size, direction, etc.), are rich with information about the social interaction. In some cases, the robot may have sensors that provide physiological responses, such as heart rate, blood pressure, body temperature, and galvanic skin response (the conductivity of the skin), which also give very useful cues about the interaction.

But while measuring the changes in heart rate and body temperature is very useful, the robot still needs to be aware of the person's face in order to "look" at it. *Social gaze*, the ability to make eye contact, is critical for normal social interpersonal interaction, so even if the robot can't understand the facial expression, it needs at least to point its head or camera to the face. Therefore, vision research is doing a great deal of work toward creating robots (and other machines) capable of finding and understanding faces as efficiently and correctly as possible.

And then there is language. The field of *natural language processing (NLP)* is one of the early components of artificial intelligence (AI) , which split off early and has been making great progress. Unfortunately for robotics, the vast majority of success in NLP has been in written language, which is why search engines and data mining are very powerful already, but robots and

HUMAN INFORMATION PROCESSING

PROXEMICS

SOCIAL GAZE

NATURAL LANGUAGE PROCESSING (NLP)

computers still can't understand what you are saying unless you say very few words they already know, very slowly and without an accent. Speech processing is a challenging field independent from NLP; the two sometimes work well together, and sometimes not, much the same as has been the case for robotics and AI. Great strides have been made in human speech processing, but largely for speech that can be recorded next to the mouth of the speaker. This works well for telephone systems, which can now detect when you are getting frustrated by endless menus ("If you'd like to learn about our special useless discount, please press 9" and the all-time favorite "Please listen to all the options, as our menu has changed"), by analyzing the qualities of emotion in speech we mentioned above. However, it would be better if robot users did not have to wear a microphone to be understood by a robot. To make this possible, research is taking place at the levels you learned about in Chapter 7: hardware, signal processing, and software.

The above just scrapes the surface of the many interesting problems in HRI. This brand-new field that brings together experts with backgrounds not only in robotics but also in psychology, cognitive science, communications, social science, and neuroscience, among others, will be great to watch as it develops.

22.6   Service, Assistive and Rehabilitation Robotics

SERVICE ROBOTS   The twenty first century will witness a great spectrum of *service robots*, machines used in a variety of domains of human life. Some of the more mundane ones, which include window washers and fetch-and-carry delivery systems, are already deployed in controlled environments such as factories and warehouses. The more interesting systems, aiming at daily use in hospitals, schools, and eventually homes, are almost within reach. The Roomba, shown earlier, is an example of one such system where many are expected to follow.

ASSISTIVE ROBOTICS   *Assistive robotics* refers to robot systems capable of helping people with special needs, such as individuals convalescing from an illness, needing rehabilitation following an accident or trauma, learning or training in a special setting, or aging at home or in a managed care facility. For example, Pearl the Nursebot (shown in Figure 22.10), designed at Carnegie Mellon University, roamed around an elder care home and helped people find their way to the dining room, the TV room, and other important places. Did you know that older people who have a pet or even a plant tend to live longer and report that they are happier, but often those pets and plants go unfed? Perhaps

Figure 22.10 Nursebot, developed at Carnegie Mellon University. (Photo courtesy of Carnegie Mellon University)

robots can be effective companions to people, and make them happy and live longer without requiring food in return. Who knows?

REHABILITATION ROBOTS

There are many different ways in which people can receive assistance. *Rehabilitation robots* provide hands-on help by moving the parts of the body of the patient in order to guide prescribed exercises and recovery. Figure 22.11 shows a rehabilitation robot and computer system; the robot moves the patient's arm while the screen provides instructions.

SOCIALLY ASSISTIVE ROBOTICS

Another way to help people involved in rehabilitation is not to guide their arms and legs, but to provide social interaction through HRI. *Socially assistive robotics* studies robots that are capable of monitoring their users and providing coaching, motivation, and encouragement without physical contact. Figure 22.12 shows Thera, a robot developed at the University of Southern California, which helped stroke patients perform their exercises. During the two to three months immediately after a stroke, the patient has by far the best chance to avoid permanent disability, but only if she properly performs very carefully designed exercises. Because over 750,000 people have strokes each year in the United States alone, and that number is expected to double in the next twenty years with the growing elderly population, robots can play a key role in helping a lot of people recover from stroke.

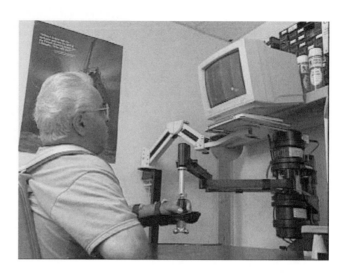

Figure 22.11   A rehabilitation robot in action. (Photo courtesy Dr. Hermano Krebs)

Figure 22.12   A socially assistive robot helping a stroke patient. (Photo courtesy of the author)

Figure 22.13 Another socially assistive robot.

Socially assistive robotics is focused on developing robots for a variety of users beyond stroke patients. For example, robots are being used as tools for studying human social behavior and various social anomalies, ranging from pathological lying to extreme shyness. Such robots can also be used to help children with autism spectrum disorder (ASD), who are particularly interested in machines and are often very talented, but have serious social and developmental problems. Socially assistive robots may also be able to help kids with attention deficit and hyperactivity disorder (ADHD). By creating robots that can be intelligent and autonomous companions and helpers, we can discover entirely new areas where such robots can improve human quality of life.

22.7 Educational Robotics

What better place to improve than in school? As you saw in Chapter 21, robots can learn. More importantly, they can teach, and serve as excellent tools for teaching. Children and people of all ages and cultures love to play with robots, program them, and push them to their limits. As you no doubt noticed, understanding how robots work and programming them to do what you want is far from simple, and as you learned how to do it, you also learned a great deal more. Working with robots provides experiential, hands-on education which most people who try it find very interesting and motivating. Imagine working with the robot in figure 22.14; would that make

Figure 22.14   A robot can be an interesting learning tool and buddy. (Photo courtesy of the author)

class more fun? Few people are bored by robots they can program and control, because the challenge is open and endless.

People love challenges, so robotics has been shown to be a superb tool for hands-on learning, not only of robotics itself (as we have done in this book) but also of general topics in science, technology, engineering, and math, also STEM TOPICS called *STEM topics*. As it happens, STEM topics are not as popular with children and students as they should be, considering what great jobs, careers, and incomes are available in those areas. This creates an undersupply of trained people to take up those great jobs, so increasing attention has been paid to developing innovative tools for improved teaching of STEM topics. Robotics is at the top of the list.

In addition to robotics courses in and after school and as part of university curricula, a great many robotics contests have been popping up. These include FIRST Robotics, LEGO Mindstorms (shown in figure 22.15), Botball, Micromouse, and many others.

Because robotics is a great field to learn about and a great tool for learning EDUCATIONAL in general, the field of *educational robotics*, which creates such learning ROBOTICS materials, is expected to grow in the future. We can look forward to all ele-

Figure 22.15   Some creative LEGO robots being put to good use. (Photo courtesy of Jake Ingman)

mentary students having an opportunity to explore robotics in order to get excited about learning.

22.8   Ethical Implications

Before we finish this book, we need to think about one very important aspect of any rapidly changing technology, including robotics. The media, in particular the movies, have often portrayed robots as evil. Why not? After all, they make such great bad guys. In all that bad publicity, it is easy to forget the major motivations of the field of robotics, which are aimed at helping people. However, any technology has the potential to be abused, and robotics is no exception. As we develop more intelligent and capable robot systems, we need to take great care about many important ethical issues. Here are some of them:

- Safety: No robot should hurt a human being. This is Asimov's First Rule of Robotics, and it's a great one to follow. Robots must naturally be safe when they interact with people, such as in the applications we listed earlier in the chapter. But people should also be safe when they design robots, and stay away from human-killing machines. People often worry about robots "going bad." But the real concern is people going bad; *robots are only as good or bad as the people who create them.*

- Privacy: Robots that become effective at working with people and in human environments will see and hear a great deal of information that they could, potentially, pass on to somebody who should not see or hear it. Privacy has to be taken seriously when the robot is designed, not after the fact, as a result of users' complaints.

- Attachment: If robots succeed in being engaging and useful, their human users will become attached to them, and will not want to throw them out or let go of them when the lease/lifetime/warranty expires. (Consider how many people still drive very old cars in spite of the cost, or stick to old computers or obsolete technology due to habit.) But what happens when an outdated robot can no longer be maintained? Roomba users already refuse to have their Roombas replaced when they need repair, insisting on getting the same one back. What happens when the robot is much more interesting, intelligent, and engaging than the Roomba?

- Trust and autonomy: Choosing to use a robot implies trusting the technology. What if the robot, out of uncertainty, an error, or simply its own best judgment, acts in a way that the user finds unacceptable and therefore untrustworthy? Nobody would want a pushy or bossy robot, but presumably everybody would want a robot that can save their life. How does a robot earn and keep the user's trust? How much autonomy should the robot have, and who is really in charge?

- Informed use: Users of robots need to be aware of all of the issues above and any others that are involved in the system they are using. As technology becomes more complex, it is increasingly difficult to know everything about the system (Do you know all the features of your phone?), yet those features may have very serious implications for safety, privacy, attachment, and trust. Designing systems that are easy to use and allow the user to know them completely is a major open challenge in technology.

In Summary

The future of robotics is in your hands! Go do something great, but remember that innovation and discovery involve risks. Think about the potential uses of what you are developing, beyond the obvious, and to the unintended. Be careful, be smart, and do some good with robotics.

Food for Thought

- While the potential for robots to help people is vast, some people believe that instead of creating technology that replaces human labor, we should focus on training people as providers of those services. With the exception of jobs that are inherently dangerous or undesirable, it is important to be able to properly justify robots in roles that could be filled by people, if such people were available. How do you think this should be justified? Are robots taking these roles away from people or are they filling otherwise unfilled roles? What if those roles are unfilled now, but once robots fill them, they could never again be filled by people? These are just some of the ethical and economic questions that have been posed. How would you answer them?

- Some philosophers have argued that once robots become sufficiently intelligent and self-aware, they will be so much like people that it will be unethical to have them working for us, as it will create a race of slaves. What do you think about that?

Looking for More?

Well, there is no more in this book, except for the glossary of terms and the index.

If you haven't yet tried all the robot programming exercises in the the Robotics Primer Workbook, you should:
http://roboticsprimer.sourceforge.net/workbook/.
Since the workbook on the web is on a public site, a growing community will be able to add new exercises and solutions, so it's worth returning to the site to both contribute and learn more.

There are numerous books and articles, popular and scientific, that you can read about the future robotics. And of course there is science fiction, the field that produced the term "robot" (remember Chapter 1?) and all manner of robot characters.

The best place to start in science fiction about robotics is with Isaac Asimov, who wrote a truly impressive number of books. *I, Robot* is a natural first read;

but read the book, don't just see the movie, since the former is incomparably better than the latter.

Based on what you have learned so far, you should be able to weed out the wheat from the chaff in your future readings about robotics. Best of luck, and enjoy!

Bibliography

Edwin A. Abbott. *Flatland: A Romance of Many Dimensions*. Dover Publications, Mineola, NY, 1952.

Harold Abelson and Gerald Jay Sussman. *Structure and Interpretation of Computer Programs*. The MIT Press, Cambridge, MA, 1996.

Ronald Arkin. *Behavior-Based Robotics*. The MIT Press, Cambridge, MA, May 1998.

W. R. Ashby. *Introduction to Cybernetics*. Chapman and Hall, London, England, 1956.

Isaac Asimov. *I, Robot*. Gnome Press, United States of America, 1950.

Dana Ballard and Christopher Brown. *Computer Vision*. Prentice-Hall Inc., Englewood Cliffs, NJ, May 1982.

R. Beckers, O. E. Holland, and J_L Deneubourg. From local actions to global tasks: Stigmergy in collective robotics. In R. Brooks and P. Maes, editors, *Artificial Life IV*, pages 181–189. MIT Press, Cambridge, MA, 1994.

George Bekey. *Autonomous Robots*. The MIT Press, Cambridge, MA, 2005.

Rodney Brooks. *Cambrian Intelligence*. The MIT Press, Cambridge, MA, July 1999.

H. Choset, K. M. Lynch, S. Hutchinson, G. Kantor, W. Burgard, L. E. Kavraki, and S. Thrun. *Principles of Robot Motion: Theory, Algorithms, and Implementations*. The MIT Press, Cambridge, MA, June 2005.

H. Choset, K. M. Lynch, S. Hutchinson, G. Kantor, W. Burgard, L. E. Kavraki, and S. Thrun. *Principles of Robot Motion: Theory, Algorithms, and Implementations*. The MIT Press, Cambridge, MA, June 2005.

Jonathan Connell and Sridhar Mahadevan, editors. *Robot Learning*. Kluwer, Boston, MA, June 1993.

Thomas Corman, Charles Leiserson, and Ronald Rivest. *Introduction to Algorithms*. The MIT Press, Cambridge, MA, September 2001.

J. J. D'Azzo and C. Houpis. *Linear Control System Analysis and Design: Conventional and Modern*. McGraw-Hill, New York, NY, 4 edition, 1995.

J. L. Deneubourg, S. Goss, N. R. Franks, A. Sendova-Franks, C. Detrain, and L. Chretien. The dynamics of collective sorting: Robot-like ants and ant-like robots. In J-A Meyer and S. Wilson, editors, *Simulation of Adaptive Behaviour: from animals to animats*, pages 356–365. MIT Press, Cambridge, MA, 1990.

Richard C. Dorf and Robert H. Bishop. *Modern Control Systems*. Prentice-Hall, Englewood Cliffs, NJ, 10 edition, 2004.

Hugh Durrant-Whyte and John Leonard. *Directed Sonar Sensing for Mobile Robot Navigation*. Springer, Dordrecht, The Netherlands, May 1992.

H. R. Everett. *Sensors for Mobile Robots: Theory and Applications*. AK Peters Ltd., Natick, MA, 1995.

James Jerome Gibson. *The Perception of the Visual World*. Houghton Mifflin, Boston, MA, 1950.

Clive Gifford. How to build a robot, 2001.

Madan M. Gupta and Naresh K. Sinha, editors. *Intelligent Control Systems, Theory and Applications*. IEEE Press, New York, 1995.

Simon Haykin and Barry Van Veen. *Signals and Systems*. Wiley & Sons, New York, NY, 1999.

Berthold Klaus Paul Horn. *Robot Vision*. The MIT Press, Cambridge, MA, March 1986.

Paul Horowitz and Winfield Hill. *The Art of Electronics*. Cambridge University Press, Cambridge, MA, 2 edition, July 1989.

B. C. Kuo. *Automatic Control Systems*. Prentice-Hall, Englewood Cliffs, NJ, 1962.

Steven M. LaValle. *Planning Algorithms*. Cambridge University Press, Cambridge, MA, May 2006.

Steven Levy. *Artificial Life*. Random House Value Publishing, New York, NY, May 1994.

Fred Martin. *Robotic Explorations: A Hands-on Introduction to Engineering*. Prentice Hall, Englewood Cliffs, NJ, 1 edition, December 2000.

Maja J Matarić. A distributed model for mobile robot environment-learning and navigation. Master's thesis, Massachusetts Institue of Technology, Cambridge, MA, January 1990.

Maja J Matarić. Integration of representation into goal-driven behavior-based robots. *IEEE Transactions on Robotics and Automation*, 8(3):304–312, June 1992.

Tom Mitchell. *Machine Learning*. McGraw-Hill Education, New York, NY, October 1997.

Gordon E. Moore. Cramming more components onto integrated circuits. 38(8):114–117, April 1965.

Robyn Murphy. *Introduction to AI Robotics*. The MIT Press, Cambridge, MA, 2000.

Alan V. Oppenheim, Alan S. Willsky, and Nawab S. Hamid. *Signals and Systems.* Prentice-Hall, Englewood Cliffs, NJ, August 1996.

Paolo Pirjanian. *Multiple Objective Action Selection & Behavior Fusion using Voting.* PhD thesis, Aalborg University, Denmark, April 1998.

Julio Rosenblatt. *DAMN: A Distributed Architecture for Mobile Navigation.* PhD thesis, Carnegie Mellon University, Pittsburgh, PA, 1997.

Stuart Russell and Peter Norvig. *Artificial Intelligence, A Modern Approach.* Prentice-Hall, Englewood Cliffs, NJ, December 2002.

L. Sciavicco and B. Siciliano. *Modeling and Control of Robot Manipulators.* Springer, Great Britian, 2 edition, 2001.

Roland Siegwart and Illah Nourbakhsh. *Autonomous Mobile Robots.* The MIT Press, Cambridge, MA, April 2004.

Michael Sipser. *Introduction to the Theory of Computation.* PWS Publishing Company, Boston, MA, 1996.

Mark Spong, Seth Hutchinson, and M. Vidyasagar. *Robot Modeling and Control.* John Wiley and Sons Inc., United States of America, November 2005.

Rich Sutton and Andy Barto. *Reinforcement Learning: An Introduction.* The MIT Press, Cambridge, MA, 1998.

Ellen Thro. Robotics, the marriage of computers and machines, 1993.

Sebastian Thrun, Wolfram Burgard, and Dieter Fox. *Probabilistic Robotics.* The MIT Press, Cambridge, MA, September 2005.

W. Grey Walter. An imitation of life. 182(5):42–45, 1950.

W. Grey Walter. A machince that learns. 185(2):60–63, 1951.

Norbert Wiener. *Cybernetics or Control and Communication in the Animal and the Machine.* The MIT Press, Cambridge, MA, 2 edition, 1965.

E. O. Wilson. *Ants.* Belknap Press, Cambridge, MA, March 1998.

John Yen and Reza Langari. *Fuzzy Logic: Intelligence, Control and Information.* Prentice-Hall, Englewood, CA, November 1998.

Glossary

Achievement goal A state the system tries to reach, such as the end of a maze.

Actuator The mechanism that enables an effector to execute an action or movement.

Actuator uncertainty Not being able to know the exact outcome of an action ahead of time, even for a simple action such as "go forward three feet."

Amplifying Something that causes an increase in size, volume, or significance.

Artificial intelligence (AI) The field that studies how intelligent machines (should) think. AI was officially "born" in 1956 at a conference held at Dartmouth University, in Hanover, New Hampshire.

Assistive robotics Robot systems capable of helping people with special needs, such as individuals convalescing from an illness, rehabilitating from an accident or trauma, learning or training in a special setting, or aging at home or in a managed care facility.

Autonomy The ability to make one's own decisions and act on them.

Behavior coordination Deciding what behavior or set of behaviors should be executed at a given time.

Bit The fundamental unit of information, which has two possible values: the binary digits 0 and 1; the word comes from "binary digit."

Bottom-up Progression from the simpler to the more complex.

Broadcast communication Sending a message to everyone on the communication channel.

Calibration The process of adjusting a mechanism to maximize its performance (accuracy, range, etc.).

Cartesian robots Robots that are similar in principle to Cartesian plotter printers, and are usually used for high-precision assembly tasks.

Command arbitration The process of selecting one action or behavior from multiple possibilities.

Continuous state State that is expressed as a continuous function.

Control architecture A set of guiding principles and constraints for organizing a robot's control system.

Cooperation Joint action with a mutual benefit.

Coordination Arranging things in some kind of order (*ord* and *ordin* mean "order" in Latin).

Cybernetics A field of study that was inspired by biological systems, from the level of neurons (nerve cells) to the level of behavior, and tried to implement similar principles in simple robots, using methods from control theory.

Direct current (DC) motor A motor that converts electrical energy into mechanical energy.

Damping The process of systematically decreasing oscillations.

Data association problem The problem of uniquely associating the sensed data with absolute ground truth.

Degrees of freedom The dimensions in which a manipulator can move.

Deliberative control Type of control that looks into the future, so it works on a long time-scale.

Demodulator A mechanism that is tuned to the particular frequency of the modulation, so it can can be decoded.

Desired state The state the system wants to be in, also called the goal state.

Detector A mechanism that perceives (detects) a property to be measured.

Differential drive The ability to drive wheels separately and independently, through the use of separate motors.

Diffuse reflection Light that penetrates into the object, is absorbed, and then comes back out.

Discrete state Different and separate state of a system (from the Latin *discretus* meaning "separate"), such as down, blue, red.

Dynamics The study of the effects of forces on the motion of objects.

Effector Any device on a robot that has an effect (impact, influence) on the environment.

Embodiment Having a physical body.

Emitter A mechanism that produces (emits) a signal.

Error The difference between a desired value and the measured value.

Exo-skeleton The outer skeleton (from the Greek *ex* meaning "outside").

External state The state of the world, as the robot can perceive it

Internal state The state of the robot, as the robot can perceive it.

Exteroception The process of sensing the world around the robot, not including sensing the robot itself.

Feedback The information that is sent back, literally "fed back", into the system's controller.

Fixed hierarchies Hierarchies whose ordering does not change, like royal families where the order of power is determined by heredity.

Focus A point toward which light rays are made to converge.

Foraging The process of finding and collecting something items from some specified area, such as harvesting a crop or removing trash from a stretch of highway.

Fusion The process of combining multiple possible possibilities into a single result.

Ganged gears Gears positioned next to each other so their teeth mesh and, together, they produce a change of speed or torque of the attached mechanism. Also called gears in series.

Grasp points Locations where the fingers or grippers should be placed in order to best grasp an object. Grasp points are computed relative to the center of gravity, friction, location of obstacles, etc.

Goal state The state the system wants to be in, also called the desired state.

Haptics The study of the sense of touch.

Heuristics Rules of thumb that help guide and hopefully speed up the search.

Hierarchies Groups or organizations that are ordered by power (from the Greek *hierarkh* meaning "high priest.")

Holonomic Being able to control all available degrees of freedom (DOF).

Human-robot interaction (HRI) A new field of robotics focusing on the challenges of perceiving and understanding human behavior in real time (who is that talking to me, what is she saying, is she happy or sad or giddy or mad, is she getting closer or moving away?), responding in real-time (what should I say? what should I do?), and doing so in a socially appropriate and natural way that engages the human participant.

Hybrid control Approaches to control that combine the long time-scale of deliberative control and the short time-scale of reactive control, with some cleverness in between.

Image plane The projection of the world onto the camera image; it corresponds to the retina of the biological eye.

Inhibitory connection A connection in which the stronger the sensory input, the weaker the motor output.

Inverse pendulum problem Controlling a system that is like an upside-down pendulum, such as balancing in one-legged robots.

Joint limit The extreme of how far a joint can move.

Kinematics The correspondence between actuator motion and the resulting effector motion.

Lens The structure that refracts the light to focus it on the retina or image plane.

Level of abstraction Level of detail.

Life-long learning The idea of having a robot learn continually as long as it is functional.

Linear actuator An actuator that provides linear movement, such as getting longer or shorter.

Maintenance goal A goal that requires ongoing active work on the part of the system, such as "keeping away from obstacles."

Manipulation Any goal-driven movement of a manipulator.

Manipulator A robotic manipulator is an effector.

Manipulator links Individual components of the manipulator.

Model-based vision An approach to machine vision which uses models of objects (prior information or knowledge about those objects) represented and stored in a way that allows comparison and recognition.

Natural language processing (NLP) The field that studies understanding of written and spoken language. A part of Artificial Intelligence.

Negative feedback Feedback that gets smaller in response to the input, resulting in damping.

Non-holonomic Not being able to control all available degrees of freedom (DOF).

Odometry Keeping track/measuring how far one has gone (from the Greek *hodos* meaning "journey" and *metros* meaning "measure").

Optimization The process of improving a solution to a problem by finding a better one.

Optimizing search The process of looking for multiple solutions in order to select the best one.

Passive actuation Utilizing potential energy in the mechanics of the effector and its interaction with the environment to move the actuator, instead of using active power consumption.

Perception See Sensing.

Photocell A cell that is sensitive to the amount of light that hits it.

Photosensitive elements Elements that are sensitive to the light, such as rods and cones in the biological eye.

Pixel A basic element of the image on the camera lens, computer, or TV screen.

Polarized light The light whose waves travel only in a particular direction, along a particular plane.

Polarizing filter A filter that lets through only the light waves with a particular direction.

Polygon of support The area covered by the ground points of a legged object or robot.

Position control Controlling a motor so as to track the desired position at all times.

Positive feedback Feedback that gets larger in response to the input, resulting in amplification.

Proprioception The process of sensing the state of one's own body.

Pulse-width modulation Determining the duration of the signal based on the width (duration) of the pulse.

Putting through Having the learner experience the task directly, through its own sensors.

Quadrature shaft encoding A mechanism for detecting and measuring direction of rotation.

Reactive systems Systems that do not use any internal representations of the environment, and do not look ahead at the possible outcomes of their actions; they operate on a short time-scale and react to the current sensory information.

Reactive control A means of controlling robots using a collection of prioritized rules, without the use of persistent state or representation.

Reconfigurable robots Robots that have modules or components that can come together in various ways, creating different shapes and differently shaped bodies.

Redundancy The repetition of capabilities within a system.

Redundant Robot or actuator that has more ways of control than the degrees of freedom it has to control.

Rehabilitation robots Robots that provide hands-on help by moving the parts of the body of the patient in order to guide prescribed exercises and recovery.

Representation The form in which information is stored or encoded in the robot.

Resolution The process of separating or breaking something into its constituent parts.

Robustness The ability to resist failure.

Rotational DOF The possible ways in which a body can rotate (turn).

Segmentation The process of dividing up or organizing the image into parts that correspond to continuous objects.

Sensing The process of receiving information about the world through sensors.

Sensor fusion Combining multiple sensors to get better information about the world.

Sensor preprocessing Processing that comes before anything else can be done in terms of using the data to make decisions and/or take action.

Sensor-actuator networks Groups of mobile sensors in the environment that can communicate with each other, usually through wireless radio, and can move around.

Servo motor A motors that can turn its shaft to a specific position.

Signal to symbol The problem of going from the output of a sensor to an intelligent response.

Situated To exist in a complex word and to interact with it.

Situated automata Abstract (not physical) computing machines with particular mathematical properties that interact with their (abstract) environments.

Smoothing Application of a mathematical procedure called convolution, which finds and eliminates isolated peaks in the signal/data.

Social gaze Making eye contact in social and inter-personal interaction.

Socially assistive robots Robots that are capable of monitoring their users and providing training, coaching, motivation, and encouragement without physical contact.

Sonar Ultrasound (from SOund Navigation And Ranging).

Space All possible values or variations of something.

Spatial credit assignment The general problem of assigning the credit or blame to actions taken by members of a team.

Specular reflection The reflection from the outer surface of the object. The sound wave that travels from the emitter bounces off multiple surfaces in the environment before returning to the detector.

State The description of a system. A general notion from physics that is borrowed by robotics (and by computer science and AI, among other fields).

State estimation The process of estimating the state of a system from measurements.

State space Consists of all possible states a system can be in.

Static stability Being able to stand without having to perform active control to prevent from falling over.

Statically stable walking A robot that can walk while staying balanced at all times.

Statistical neural networks A set of formal mathematical techniques from statistics and probability applied to neural networks.

Stereo vision The ability to use the combined points of view from the two eyes or camers to reconstruct three-dimensional solid objects and to perceive depth. Formally called binocular stereopsis.

Stigmergy The form of communication in which information is conveyed through changing the environment.

Supervised learning An approach to learning in which an external teacher provides the answer or at least tells the robot what it did wrong.

Tele-operated Operated from afar (from the Greek *tele* meaning "far").

Temporal credit assignment The general problem of assigning the credit or blame to actions taken over time.

Time series A sequence of data provided over time.

Topological map A map represented as a collection of landmarks connected with links.

Torque control An approach to control which tracks the desired torque at all times, regardless of its specific motor shaft position.

Trajectory and motion planning A computationally complex approach that which involves searching through all possible trajectories and evaluating them, in order to find one that will satisfy the requirements.

Transducer A device that transforms one form of energy into another.

Translational DOF DOF that allows the body to translate, meaning move without turning (rotation).

Tripod gait A statically stable gait in which three legs stay on the ground, forming a tripod, while the other three lift and move.

Uncertainty The inability to be certain, to know for sure, about the state of itself and its environment, in order to take absolutely optimal actions at all times.

Uncontrollable DOF The DOF that are not controllable.

Universal plan A set of all possible plans for all initial states and all goals within the state space of a particular system.

Vestibular occular reflex (VOR) The reflex that keeps the eyes fixed while the head is moving.

Wikipedia A great ever-growing free encyclopedia found on the Web at http://en.wikipedia.org/wiki/ that has a lot of information about everything, including various topics covered in this book.

Index